DNA-RNA

Research for Health and Happiness

JOSE MORALES DORTA, PHD

BALBOA
PRESS
A DIVISION OF HAY HOUSE

Copyright © 2018 Jose Morales Dorta, PhD.

All rights reserved. No part of this book may be used or reproduced by any means, graphic, electronic, or mechanical, including photocopying, recording, taping or by any information storage retrieval system without the written permission of the author except in the case of brief quotations embodied in critical articles and reviews.

Balboa Press books may be ordered through booksellers or by contacting:

Balboa Press
A Division of Hay House
1663 Liberty Drive
Bloomington, IN 47403
www.balboapress.com
1 (877) 407-4847

Because of the dynamic nature of the Internet, any web addresses or links contained in this book may have changed since publication and may no longer be valid. The views expressed in this work are solely those of the author and do not necessarily reflect the views of the publisher, and the publisher hereby disclaims any responsibility for them.

The author of this book does not dispense medical advice or prescribe the use of any technique as a form of treatment for physical, emotional, or medical problems without the advice of a physician, either directly or indirectly. The intent of the author is only to offer information of a general nature to help you in your quest for emotional and spiritual well-being. In the event you use any of the information in this book for yourself, which is your constitutional right, the author and the publisher assume no responsibility for your actions.

The information in this book is not intended to be used for medical or psychiatric diagnosis of any person. The illustrations included are approximations of the subject matter for ease of understanding.

Any people depicted in stock imagery provided by Thinkstock are models, and such images are being used for illustrative purposes only. Certain stock imagery © Thinkstock.

Print information available on the last page.

ISBN: 978-1-5043-6154-5 (sc)
ISBN: 978-1-5043-6156-9 (hc)
ISBN: 978-1-5043-6155-2 (e)

Library of Congress Control Number: 2016910943

Balboa Press rev. date: 07/07/2016

Contents

I. The DNA Double Helix 1
    How Nucleic Acids Become Amino Acids 6
    Transcription Follows Replication 11
    The First MicroRNA 13
    My Interest in RNA 14

II. Fighting Diseases 17
    Mouse Brain Tissue 18
    Genome-Wide Association Studies 20
    C. Venter Challenged HGP 21
    Francis Crick and RNA 23
    After the HGP Completion in 2003 25
    DNA Packing 28
    A Seeming Contradiction 29
    DNA Methylation 31
    Chronic Stress and Your Brain 32
    Body Energy and Wisdom 34
    The Inquisitive Brain 38

III. Epigenetics 41
    More on Epigenetics 43
    Posttranslation Modifications 45

Maternal Nurturing and DNA Methylation 46
Cancer and Epigenetics 47
Genetic Markers 48
The DNA Structures in the Brains of Our Scientists 50
Watson and Crick Parade in the Canyon of Heroes 52
The Very Old DNA and RNA Molecules 53
Introns and Exons 54

IV. Thalassemia, Hematopoiesis, and Other Related Disorders .. 56
The Pasteur Institute 57
Harry Noller and RNA 60
The Bubble Boy, David 63

V. Learning from Sequenced Genomes 65
DNA, SNPs, and Diseases 66
Genome Surprises 67
The Last Ten Years in Genome Sequencing 69

VI. The Watson, Crick, and Venter Genomes 71
Gene Targeting Is Not Sci-Fi 72
The Science King of the Twenty-First Century 73

VII. Research in RNA, Amino Acids, and Proteins 76
G Protein Receptors 79
A Personal Encounter with Histamine 82
Histamine as a Chemical Messenger 84

VIII. E. Kandel's Research 86
Alois Alzheimer and Spaghetti Strings 88
Following on Proteins 89
From Mind Speculation to Brain Research 90

Our Basic Carbon Composition 92
From Transcription to Proteins 94
Proteins, Memories, and Eric Kandel 96
More on Memories and Learning 98
Memories, the Brain, and Learning 100
CREB Protein 101

IX. Nobel Laureates Advocating RNA 104
Previously Unknown RNA Functions 107
A Little Bit of History 109
An Established Fact—an RNA World 110
Dr. Tom Misteli from the NCI and DNA 112
MicroRNA's Regulatory Role 113
Unbelievably Naïve 114
Here and Now with Amino Acids and Proteins 115
St. Augustine and Darwin Join Our Group 116
Stress and ADHD in Pregnant Women 117
Daydreaming 118
The Brain—Just 2 Percent 119
Raichle, Shulman and the Default Network Mode 120
Dr. Marcus Raichle's Opinion 122
Pete, the Autistic Boy in my Office 123
Isolation and Loneliness in Schizophrenic and Autistic
Individuals 125

X. Your Brain and Plasticity 127
A Nondogmatic Believer, J. Cotard 128
1876 Was a Long Time Ago 129
P. Bach-y-Rita and Plasticity 130
Self-Isolation, Not Loneliness 131
The Hunt for Diseases and Genome Sequencing 133

Mobility within the Genome 134
Reinforcing Erroneous Assumptions 135

XI. A Bug inside My Brain? 138
You Are Unique in Our Solar System 140
Committed and Enthusiastic, but Naive 141
Protein Encoding Problems 142
David Lewis and Schizophrenia 144

XII. Atoms and Molecules in Action 146
Covalent Bonds 147
Notes from My Teacher 151
The DNA Replication Process 152
Your Chemistry: DNA 153
Tiny Particles inside the Atom 154

XIII. DNA Technology and Restriction Enzymes 158
TV Shows and DNA Fingerprinting 161
Attempts to Hide DNA Fingerprints 162
Electrophore Technology 163
Recombinant DNA 164
The Laboratory Workhorse—a Bacterium 165
Just One More Time 166
My Noxious Friend and Tenant E. Coli 168
A Research Hiatus 171
Protein and enzyme functions 171
Learning the Trade 174
A Virus Cannot Multiply Alone 175

XIV. Illness under the Electron Microscope 177
Anxiety 178

An Almond-Shaped Cluster of Brain Cells......................... 179
An Animal Cell and Amino Acids in Pictures.................... 183
Inside an Animal Cell 184
From Research to Clinical Practice 185

XV. Mary: From a Soft Couch to a Hard Chair 214

XVI. Helga, the Star of the Circus 223

The DNA Double Helix

THE DATE WAS FEBRUARY 28, 1953. There was no news coming from the Korean War that excited James D. Watson's brain cells. The day before, chemist J. Donahue from the California Institute of Technology had made a significant correction to the nucleic acid textbook Watson was relying on to construct a structural model for the DNA macromolecule. How many chains is the structure composed of: two, three, or more? Where would the adenine, thymine, cytosine, and guanine go? If I place them on the outside, hanging from the DNA backbone, can they come together to replicate themselves, creating a genetic code for all animals and plants? How can the bases A, T, C, and G be bonded to each other and pass on the genetic message to future generations? Which bases attract or repel each other? How do we go about testing our hypothesis? The question of whether protein or DNA was the hereditary material had been scientifically established at Cold Spring Harbor in New York by Alfred Hershey and Martha Chase. American and British scientists seemed to have been engaged in a race to discover the structure of the book of life. How was this book written? Was it written in a code, or did it randomly come together? Was the code a chemical or an electrical one?

There were brilliant and outstanding chemists, physicists, and biologists on both sides of the Atlantic Ocean who were very much interested in making the covers of scientific journals and major newspapers around the world. In the United States, we had world-renowned two-time Nobel Prize–winning chemist Linus Pauling. He was very much engaged in the race to bring to America the Nobel trophy. In England, there was Rosalind Franklin, a pioneer X-ray crystallographer who was very close to discovering the double helix DNA model. Not too far from Rosalind's workplace was Maurice Wilkins, who was employed at the biophysics laboratory of King's College. These two brilliant English scientists were working close to each other. However, their own bodily chemistry kept them miles apart, except when Watson and Crick—but especially Watson— came around trying to push their double-helix model.

James D. Watson, an American biologist, was desperately trying to find his place among the most prominent scientists of his time. His traveling between America and Europe put him in contact with influential scientists and scientific organizations on both sides of the ocean. His scientific curiosity, approachable personality, and perseverance landed him in a research position at the Cavendish Laboratory at Cambridge University in the United Kingdom. The maverick and future Nobel laureate J. D. Watson was blessed by sharing space in the biochemistry laboratory with Francis Crick. A WWII physicist, Crick, who was twelve years older than Watson, was working toward his PhD in biology. Watson and Crick, an American and a Briton, must have had good chemistry suitable for mutual compatible communication. Francis Crick seemed to have a gregarious and contagious mood and personality. In many ways, he was the perfect complement to Watson. Crick not only came out with brilliant ideas of his own but, for the good of both, he was always carefully attentive to Watson's observations when listening to his colleagues or attending lectures and conferences by other scientists. When something did not go right, it was expected at Cavendish for Crick's good humor to turn it into an insightful moment.

On the other hand, there were two hardheaded Britons, each one following his own research goals and thoughts. Rosalind Franklin had many years of formal training and experience over both Crick and Watson, and she wanted test results on crystallography diffraction before making any premature announcements. Cause and effect, as well as test results, were her favorite criteria for claiming a job had been done well.

The other key player in the double helix was physicist Maurice Wilkins, whose experience included working on the Manhattan Project. Whether he met and shared his work with Albert Einstein and Robert Oppenheimer is unknown. However, we can assume that Wilkins was a good team player, as he was able to survive the secrecy his job demanded.

Despite professional intrigues, he maintained friendly communication with R. Franklin and cooperated with Watson and Crick while working on the double helix. As already mentioned, America had a superstar chemist, L. Pauling at Caltech, who dictated how chemistry should be studied and practiced. Hardly anyone would challenge his arguments, as he had discovered the exact sequence in which amino acids fold up into proteins—the building blocks of our body. However, on the DNA double-helix structure, he missed the point by a very small fraction. Otherwise, he would have been a three-time Nobel laureate. Wilkins, the physicist, may have had much in common with Rosalind Franklin on research principles, but socially they must have been worlds apart. As J. D. Watson put it, when the lab day was over, Franklin dropped her lab apron and put on a stylish and distinctive evening dress and mingled with London's elite society.

However, it seems that neither Watson nor Crick would be intimidated by occasional roadblocks impeding or delaying their search for the DNA-molecule structure. The United Kingdom's Cambridge University was a gold mine of information on DNA. Both scientists complied with their superiors' orders to work on proteins at the expense of the double-helix model. But their ultimate goal, the DNA-molecule structure, never left their brains. Watson's eye caught a scientific paper reporting that his much-looked-for DNA molecule bases—adenine, thymine, guanine, and cytosine—appeared in roughly equal amounts. Watson's brain was well focused in grasping this chemical balance inside his ideal DNA molecule. But it was his ever-listening friend and colleague, Francis Crick who foresaw the beginning of the solution for the problem: the chemical attraction of each base pair toward others. Adenine would attach itself only to thymine, and guanine to cytosine. The presence of two chains, each one holding opposite bases and chemically attracting each other, could explain the structural model of the double helix and subsequent duplication.

Rosalind Franklin's X-ray photos of the molecule were a decisive factor in achieving this historic moment in human biology.

On February 28, 1953, the key features of the DNA model all fell into place. The two chains were held together by strong hydrogen bonds between the base pairs of adenine and thymine, and guanine and cytosine.[1] It all fell in place almost as Rosalind had seen it during her visit to Watson and Crick's laboratory. Subsequently they enjoyed the Nobel Prize; unfortunately Rosalind died of cancer before the award was given. No Nobel Prize is given posthumously.

On April 25, 1953, the prestigious science journal *Nature* published a one-page article with the title "Molecular Structure of Nucleic Acids: A Structure for Deoxyribose Nucleic Acid." It had been authored by J. D. Watson and F. H. C. Crick.

There it was, a single page showing the drawing of this double helix with its sugar-phosphate backbone on the outside in a twisted

---

[1] James D. Watson, *DNA* (New York: Alfred A. Knopf, 2003), 52.

fashion, holding inside the nucleic acids. How tempting it was. When I first saw it, I wondered whether it was possible that I had the book of life in front of me.

The authors had written, almost at the end of the article, "It has not escaped our notice that the specific pairing we have postulated suggests a possible copying mechanism for the genetic material." Did this refer to a possible mechanism for copying the genetic material? Did it mean that there is a genetic code for me to find out how I was made? The basic elements composing the human body, including protein structures, had already been shown to us by L. Pauling. With this single page, life took a turn away from how it had been taught for centuries. From now on, the focus of biology would be to understand the working dynamics of the double helix and the nucleic acids forming it. That single page provoked more questions than any other discovery in the twentieth century, according to most biologists and scientists in general. The world had hardly had time to digest and accommodate itself to Darwin's *On the Origin of Species* before it was frightened by a code of letters—A, T, C, and G—that could open the alphabet of our book of life.

How do the double strands separate from each other for self-replication? How and where are proteins formed? Is the code for protein synthesis implied in the four letters? Francis Crick postulated the central dogma of molecular biology. How do DNA, RNA, and proteins relate to each other? Is the nucleus of the cell the command post for all cellular activities? Where in the cell are the letters synthesized? How many letters are there in the book of life?

Watson and Crick, in a one-page article published on April 25, 1953, in the journal *Nature*, inspired biologists, chemists, and physicists to engage in science 24-7, even during their dreams.

Protein synthesis kept scientists busy until they discovered ribosomes that are assembly sites for proteins. It was Paul Zamecnik and his team at Massachusetts General Hospital who further elucidated for us the role of RNA in protein formation. They confirmed Crick's adaptation theory, coining the term *transfer RNA* (tRNA). This molecule hauls amino acid groups that exactly match another RNA strand coming from the nucleus of the cell, dubbed *messenger RNA*. In 1960, messenger RNA was proven to be the true template for protein synthesis. We had not only discovered the letters of life's alphabet but also began to learn how the acids of life work. Three world-renowned American research centers—Harvard, Caltech, and Cambridge—had done the work. M. Messelson, F. Jacobs, and S. Brenner demonstrated that ribosomes existing outside the cell nucleus were, in effect, protein assembly factories.

There are many ribosomes inside a cell. They look like tiny beads set in line in the cell's cytoplasm. The inside of the cell was becoming clear for scientists looking for answers in biology. When they found the different structures and functions of various organelles, it was exciting news for scientists and the general public. We were working with the human genome, and we were excited with the discoveries of this new science, molecular biology. Our enthusiasm prompted some individuals to make premature announcements. The cell's inner organelles, their shapes and functions, were under constant microscopic scrutiny and testing. Nobel laureate Ada Etil Yonath, who participated in a meeting of the World Chemistry Congress in Puerto Rico, said that 60 percent of the ribosome is RNA. She added that RNA is a machine from before life evolved. Life grew around it. RNA was one of the first molecules.[2]

How Nucleic Acids Become Amino Acids

Unresolved as of yet, there was a nagging question: How did nucleic acids—the repeating letters A, T, C, and G, become peptide chains? In other words, there are four letters and twenty amino acids, and

---

[2] *Nature* (October 13, 2011): 56–57.

the question was regarding which and how many amino acids would fit into a letter to form a peptide chain. Once more, it was Francis Crick and his colleague at Cambridge University in 1961, Sydney Brenner, who proved that the code was a codon, also known as a triplet.

Assigning three letters to each amino acid was the solution to the problem. However, the genetic code may come in more than one codon for each amino acid. For example, the amino acid tryptophan comes in a UGG codon only, while arginine and serine may come out in six different codons. A word of caution: In the DNA sequence is thymine, which binds with adenine. In RNA, it is uracil that complements adenine. Marshall Nirenberg, working at the US National Institute of Health, participated at an international congress in biology in Moscow in 1961. Nirenberg was a young scientist hardly known outside his close friends. He was allowed to speak for about ten minutes. As with any event, attendees wanted to hear the most important personalities in the field—not an unknown neophyte. Even our J. D. Watson was chatting outside the conference room. Nirenberg reported that in his lab he had gotten a ribosome to pump out a simple protein known as phenylalanine, which comes out in codons UUU or UUC. Once more, it was Crick who arranged for Nirenberg to speak to a hall packed with every scientist that could get to Moscow the following morning. Nirenberg was later awarded the Noble Prize. Nirenberg had encouraged everyone to finish the genetic code for each amino acid.

F. Crick's central dogma regarding DNA, RNA, and proteins seemed to have been well established. From the cell nucleus, DNA is transcribed into a messenger RNA molecule that goes to a ribosome in the cytoplasm. Another RNA molecule named transfer RNA delivers amino acids to ribosomes for the synthesis of proteins. Each ribosome, acting as an assembly factory, pastes together messenger RNA and transfer RNA to form proteins. It is helpful to remember that replication occurs during the double-helix unzipping process while transcription takes place while the cell nucleus is forming single RNA strands.

Another term frequently used by most scientists when talking about these processes is *translation*. Translation is basically a process that takes place at ribosomes in the cytoplasm. Each molecular strand carries a set of letters that complement each other. For example, the messenger RNA triplet CAA will recognize and chemically attract the corresponding triplet GUU in the transfer RNA. Ribosomes will be in charge of pasting together these two molecules, thus forming peptide chains, which are known to us as proteins. The shape of the chain will determine its function, and some peptide chains are called enzymes. Enzymes are very busy catalysts in our bodies. They are always provoking chemical changes without us ever knowing anything about it. As soon as I put a teaspoon full of ice cream in my mouth, enzymes begin to do their work. Enzymes do their work inside the cell as well outside around the cell. (I will address ribonucleic acid or RNA later on in this essay, because strands of this acid seem to be the workhorse in protein forming and research tools.) Recently the traditional Watson-Crick central dogma for protein synthesis has been under review.

I will be going back and forth in this essay, trying to put together Watson and Crick's scientific success story. After Watson and Crick proposed the double-helix DNA structure and it was approved of by the majority of scientists in the field, several questions began popping up in my brain all the time: Where and how are the building blocks of our body (protein) synthesized? What roles does ribonucleic acid play in forming life? Does it compete with DNA in protein synthesis? How are ribonucleic acids involved in the complex process of peptide chains? I have already provided you some of these answers, but at the time, it was mind-boggling for all involved—especially, for Watson and Crick. It was the good-humored F.

Crick who once again suggested the central dogma regarding the relationship between DNA, RNA, and proteins. The existence of DNA polymerases, the enzymes that unzip the double helix, was well established. Actually the protein helicase does the unzipping. Soon RNA polymerase (an enzyme) came to take its corresponding place along with the DNA polymerases. The nucleus of the cell was the command center, sending out messages and ordering cells organelles what to do. However, it was RNA's multiple forms that moved around strands to assemble proteins at ribosome sites. The template for protein synthesis, messenger RNA, came out of the nucleus, but it was transported outside the nucleus and into the cytoplasm to be matched with another RNA molecule carrying an amino acid destined for a ribosome. As we indicated above, translation takes place at this level of protein synthesis. This is also the place and process where most problems take place.

Translation, whether we are referring to chemically matching nucleic acids or language translation, seems to create headaches for many of us. It is hard to find and match words or phrases that exactly convey the same meaning and feelings in any two languages. Translation in biology is an issue of RNA molecules. During the replication process, the DNA polymerase (helicase enzyme) unzips the chain of double-base pairs, and it recreates itself instantaneously by adding the necessary complementary nucleotides to each strand. The original strand of letters—for example, A for adenine, T for thymine, G for guanine, and C for cytosine—will be followed by more letters, nucleotides, in a sugar-phosphate backbone running in the opposite direction, thus forming a double strand. The nucleotides will chemically attract each other, forming strong covalent bonds. They share electrons. You may have a strand of TGACGTTAG, with all these bases being held together or attached to a sugar-phosphate backbone. In this hypothetical case, the complementary bases will be as follows: ACTGCAATC. T and A bond together first, followed by the rest of letters running antiparallel to each other. This is DNA replication. During genome sequencing, if you know one strand, the accompanying strand is not hard to guess. Your genes—the genetic material inside your cells as well as every cell in all living things on planet Earth—are made of DNA. Each strand of DNA consists of many polynucleotides and is bonded together by hydrogen bonds. Although for illustrative purposes, we tend to draw and describe each DNA strand as being in a straight and bar-like structure, DNA is a three-dimensional structure in which each strand coils around the other in a spiral fashion. You can see the length and width of the strand in most books, but the descriptions require imagination for completion.

I want to clarify that the word *nucleotide* refers not only to the four nucleic acid bases A, T, G, and C. Generally it is used to designate the bases with their sugar-phosphate group attached to them. Our DNA is well protected inside the nucleus of the cell. It is a tightly controlled structure. It needs to be wrapped that way because it needs to protect itself from invaders inside the cell cytoplasm as well as extracellular intrusion and interference.

Recently, the field of genetics as well as post-birth causal factors have indicated risks and threats to genes and possible diseases. Before I proceed with this essay on nucleic acids, I would like to remind you that the twenty amino acids needed in our organism are polymers for the synthesis of proteins. They possess an amino acid group—namely NH_2 at one end and a carboxyl acid group, COOH, at its other end. Both groups are bonded to a single carbon atom known to us as the alpha carbon. There is a side chain linked to the alpha carbon, altering the shape of the molecule and thus changing its chemical function.[3] There are more than twenty amino acids found in nature, but only twenty are needed in your body. They duplicate

---

[3] Alberts et al., *Essential Cell Biology* (New York: Garland Publishing Inc., 1998), 60.

themselves over and over again in proteins in bacteria, plants, and in a privileged civilized human being like you and me. Why nature chose only twenty amino acids to be the building blocks of protein is another question to be answered in the future. Also, there are about ninety-two elements that commonly appear in nature, but less than a third of those are used by life on Earth. Our bodies are always going through actions and reactions. The body is a marvelous and sophisticated chemical engine. Have you wondered how DNA and RNA pair together? How about mitochondria—organelles outside the nucleus—were they captured by the nucleus to do its work too?

Transcription Follows Replication

Following replication, there is transcription. With the help of RNA polymerase, the DNA molecule transcribes and forms messenger RNA, transfer RNA, interference RNA, and other ribonucleic acids strands. For the sake of clarity, when you replicate a page or paragraph of a book as a homework assignment, it most probably will come out without any mistake or error. But try to translate the same page from English to French, Spanish, or Chinese, and in all probability, your professor or teacher will find room for improvement. If you are translating poetry, it might prove even harder to make perfect translation. As I said earlier, translation is an issue of RNA molecules. RNA is not a faithful translator for reproductions of DNA genetic instructions. Watson, Crick, and their colleagues established what they termed the *central dogma*, which states that the chemical letters A T C and G are the genetic code while the cell's workhorse, the RNA, is made in its likeness with a minor exception—the uracil nucleotide. RNA, has the responsibility of transporting molecules to ribosomes for the synthesis of proteins. These proteins are used during the making of multicellular organisms, such as giraffes, elephants, turtles, and human beings. I wonder how many spelling errors are found in each genome of these animals.

Researcher Mingiao Li, a geneticist at Penn Medical School in Philadelphia, conducted a research study of RNA in which he included blood cells from twenty-seven different persons. On average, each person has nearly four thousand genes with RNA containing misspellings not found in DNA, while RNA has more than twenty thousand different places in the genome. The most common misspelling changes occur in the A letter in DNA, which is changed to G in the RNA strand.[4] The researcher took precautions that variables, including viruses, did not invalidate or tarnish his work. We must not forget that the world of RNA is not only very old, though it is abundant. While DNA sits in the nucleus of the cell giving out orders and sending out messenger RNA for the synthesis of protein, ribonucleic acids are busy transporting and delivering polymers inside the cell. We already know that without messenger RNA, there would not be protein synthesis, and without RNA in the ribosomes, peptides would not be pumped out of this assembly factory. Besides, micro-RNA is being used not only as engineering tools to work inside the cell organelles but also to modify genes.

Do not be overly concerned by the presence of so many misspellings and changes of one letter or base for another. There is a self-correcting system inside our macro-molecule DNA. Occasional changes of letters or bases are conducive to mutations, and mutations, without changing the basic core of the cell, made you and I beautiful and intelligent human beings. Ask your mother, and she will reassure you that you are the most beautiful and intelligent individual in the universe.

---

[4] *Science News* 178, no. 12 (December 4, 2010): 17.

The First MicroRNA

The first microRNA (miRNA) was discovered in 1993; the second one was discovered at Massachusetts General Hospital in the year 2000 by Gary Ruvkum. In between these two discoveries, in 1998, perhaps the most versatile RNA molecule, interference RNA, was added to the family. Among other things, it can interfere and prevent a gene from forming proteins. I would like you to remember throughout this essay the potential power of tRNAs as well as microRNAs. The potential for use and abuse of these molecules is beginning to be understood. All genetic information encoded in the DNA of eukaryotic (of an organism with one or more cells possessing a nucleus) cell genes is transcribed by RNA polymerases. The process of transcription is dominated by proteins, but a growing number of the involved molecules are RNA.[5] Recently, a new RNA molecule dubbed *enhancer RNA* (eRNA) was discovered. It belongs to a noncoding RNA group. Enhancers are engaged in activating the transcription process. It is suspected that an enhancer RNA contributes to human disease and the evolution of human specific traits.[6] I need not name for you specific human traits.

Aside from the problem encountered by the translation process and epigenetics, I would like to briefly mention the so-called junk DNA, also known as noncoding DNA. It has been the subject of much debate and scrutiny over its usefulness in our organism. If it does not code for proteins, what is the use of having an excess of cells that may be prone to health problems and diseases? Is this junk DNA left over from our earliest physiological developmental process that needs to be modernized or recycled? Perhaps noncoding DNA is like glial cells, which are neglected by scientists because neurons have been given top priority in research laboratories.

---

[5] *Nature, Vol.* 461, no. 7261 (September 10, 2010): 186.
[6] *Nature, Vol.* 465 (May 13, 2010): 173.

An alarming 90 percent of our genome is composed of noncoding genes. A recent study done at Stanford University comparing the human genetic blueprint genome with those of other animals concluded that very little of the human genome is really necessary. Andrew Sidow, the principal researcher, claims that only about 7 percent of the human genome is similar to the DNA of other mammals. This researcher deemed that 225 million of the 3 billion letters (A, T, C, and G) that compose our DNA are necessary.[7] Not all scientists in this field concur on this estimate. During the process of transcription, when a strand of messenger RNA is being made, the noncoding genes are cut off so the messenger RNA comes out clean and safe to be transported to a ribosome, where it will be converted into a protein.

My Interest in RNA

Researchers have been interested in the RNA molecule because it takes multiple forms and because in its role as a translator it is the subject of multiple problems that may contribute to protein dysfunctions and diseases. However, before I proceed on to the genome and the genetic code, I am going to say a few things about the DNA macromolecule. DNA has the potential for change without losing the basic components of its ancestors. It allows evolutionary processes to adapt and adjust it to new environments, but it maintains the basis of its genetic code. This genetic code contains and transmits information for the purpose of creating multiple different organisms. DNA establishes amino acid sequences for protein synthesis. This great molecule created a brain capable of studying and repairing itself by protein synthesis and memory storage. (RNA molecules are prominently involved during this process). DNA molecules provide for the continuity of the species with minimum changes over a long

---

[7] *Science News* 178, no. 12 (December 4, 2010): 77.

period of time; "long period" may mean thousands of years. If we accept the Neanderthal man as part of our human ancestry lineage, as I do, it is clear we have been here for a long time. Recent discoveries by Svante Paabo from Max Planck University in Germany, as well as discoveries at Atapuerca in Spain and southern Siberia in Russia, point to the presence of intelligent human beings on dear Mother Earth a long time ago.

The DNA molecule has its own membrane within the cell cytoplasm that guarantees its integrity and identity, as if it were the command center of the cell. From this complex DNA molecule comes instructions to construct a beautiful hummingbird, a mouse, an elephant, or a human being. It has the information to replicate itself as a double helix or just as simple strands of RNA. Our genetic code—the four letters A, T, C, and G,—is in itself the great DNA molecule. The way these letters are attracted to each other is written in DNA. This molecule keeps in its nucleus the instruction for the color of your eyes, hair, and skin. To some extent, it also determines how tall or short you will grow to be. Of course, environmental factors impact your height and weight. A poor diet composed of simple carbohydrates without protein-rich beans—such as soybeans, green peas, lima beans, kidney beans, and chick peas—affordable tuna, sardines, and salmon can have a strong impact. Likewise, throwing fried pork chops and medium-cooked steaks seasoned with artificial flavors and excess salt down your throat can make it extremely hard for you to be a healthy person.

Researchers have become interested in the ribonucleic acid molecule not only because it transforms itself in multiple forms but also because in its interference form it can inhibit the formation of our own bodily tissues, including brain cells.

The microRNA hairpin molecule is also another relatively new discovery. Thanks to its small size, this molecule has been used to deliver research and therapeutic tools inside the cell. It can be used to engineer gene modification in a brain disease or any other bodily pathology. It has been used in animal models. Time, dedication, and, of course, funding will decide the future of all these new technologies. MicroRNA can be used to target genes that at a certain age of bodily development provoke a dysfunction or a specific disease. These diminutive molecules can take an amino acid chain and alter the formation and functioning of a faulty protein or gene. During the blastocyst phase of embryonic growth, they can be used to silence pathological cells. Gene silencing through RNA interference and similar microRNAs has become the tool of choice for analysis of gene function, diseases, and drug discovery. Among the smallest of these molecules besides the microRNA, we have short interfering and hairpin screening tools, which are first choice among genome researchers.

Fighting Diseases

THE NEXT CHALLENGE TO Watson and Crick's book of life is to look at how the genetic code can be translated into fighting diseases and improving human health in general. From studies of mummies and animal fossils, we have learned that some diseases have been haunting us for thousands of years. The study of nucleic acids—the base pairs of letters—in our genome can be used to point out mutations, missing segments of letters, misspellings, and repeating noncoding genes.

Why has it been so difficult to fight cancer, schizophrenia, Alzheimer's and other degenerative diseases in our organism? We have engineered animal models for laboratory work, such as mice, worms, bacteria, and fruit flies, among others. It has been a challenge for all of us who would like to see human suffering alleviated by newfound discoveries, especially after WWII. Privately funded laboratory researchers, pharmaceutical companies, and state and federal agencies have joined efforts to find a solution as soon as possible. During these intervening years, much funding was diverted to space exploration in competition with the Soviet Union. The exploration of human somatic and brain cells was lagging behind. We did not have the technology—the working tools—to observe, catalog, categorize, and analyze the behavior and role of each DNA molecule. The same work had to be done with RNA molecules. Whatever little work was done, it was done by hand.

Although the double helix and both nucleic acids were known in general, the dynamics, physiology, and chemistry of the brain were left to Sigmund Freud, C. G. Jung, theologians, and poets to speculate about. For two thousand years, our cultural and religious heritage from Greek and Roman philosophers had speculated about our illnesses—especially mental diseases. The old Greek scientific knowledge was buried under superstition and philosophical fanaticism during the middle ages. It was not until the Renaissance that scholars from Byzantium introduced it to Western Europe.

Mouse Brain Tissue

The slicing of a mouse's brain tissue in order to observe how a cell grows in a petri dish or how a virus multiplies in a bacterial environment did not receive the necessary support from the public. For lovers of gene sequence history that enjoy learning about the mind, I would like to remind you that in 1985, the longest genome that had been sequenced was a virus of 172,282 base pairs. The human genome has a little over three billion pairs. We were sleeping—and sleeping soundly. That was thirty-two years after Watson and Crick had given us the genetic code of life. Three worldwide leaders and researchers took over the audacious, if not gargantuan, proposal of sequencing the human genome. We were still guessing or speculating about how many genes were in the genome. The three visionary leaders I am referring to are James D. Watson, Francis Collins, and Craig Venter of Celera Genomics.

As late as 1988, James D. Watson was named associate director of the Office of Genome Research, part of the US National Institute of Health. He was in charge of the genome project until 1992. Francis Collins took over from Watson until its sequence was completed in

2003. Craig Venter published his human genome sequence in the journal *Science*, while the NIH genome was published in *Nature*, a British journal. By the time Watson, Collins, and Venter took upon themselves the task of sequencing human DNA, the tool that multiplies DNA fragments for analysis—PCR—was available to them. The use of restriction enzymes to splice or cut DNA strands at target sites became a very useful tool not only for reading the sequence but also for marking genes involved in inherited diseases. Some of the earliest molecular biologists and scientists working with DNA structure and its content claimed that genes made up only 3 percent of DNA in human cells. That claim was made in 1995. Today we call noncoding sections of DNA introns, while sections of coding DNA are dubbed *exons*.

The race to be the first one to sequence and read the book of life was centered in America with minor international contributions. However, as governments and scientists around the world became aware of its potential benefit not only as a wonder tool for fighting disease but also as an agricultural aid that could improve food supplies across the world, networks were established for joint research projects.

With genome sequence information, seeds can be engineered to improve crop yields, and animal growth can be accelerated. Genetic engineering can be a tool par excellence in feeding hungry people around the world in regions where climate and weather disasters make it difficult to grow enough food for daily consumption.

Additionally, as more sophisticated research tools have been developed, the so-called junk DNA has come under increasing scrutiny and analysis. These noncoding genes are now revealing their usefulness.

Jose Morales Dorta, PhD

Genome-Wide Association Studies

Genome-wide association (GWA) studies have engaged hundreds of individuals, both researchers and patients, in identifying disease markers in their DNA. The book of life's language has been deciphered, waking up the minds of many brilliant and concerned persons on planet Earth. Using the advances of restriction enzymes and recombinant DNA, we have moved forward in our attempt to bring relief to suffering and hopeless individuals worldwide. We are using genes to help patients, as in gene therapy. We have been using harnessed viruses as vectors to deliver therapeutic tools to target areas. The usefulness of the human genome sequence is not in question. Many individuals have hailed it as the greatest human achievement in history. Of course there are also alarmists who have religious, philosophical, or superstitious claims that interpret man's advances as challenges to their worldview.

Following are quotations from three different professors addressing Craig Venter's latest achievement with a synthetic bacterium: "A prosthetic genome hasten the day when life forms can be made entirely from non-living materials." Another professor wrote: "Venter's achievement would seem to extinguish the argument that life requires a special force or power to exist. This makes it one of the most important scientific achievements in the history of mankind."[8] Following that, I would like you to ponder the quotation by another professor of biomedical engineering whose words appeared on page 424 of the same journal. "The micro-organism reported by Venter's team is synthetic in the sense that its DNA is synthesized, not that a new life has been created. Its genome is a stitched copy of the DNA of an organism that existed in nature, with a few small tweaks thrown in."

You now have the opinions of three different professors, all

---

[8] *Nature, Vol.* 465, no. 7297 (May 27, 2010): 422–3.

from the same article in the same journal. One praises it, another puts it down, and another sees it as an assault to almost everything we believe in. When C. Venter appeared on TV announcing the creation of a synthetic bacterium, he did not seem to claim to have formed new life. I understood it as a brilliant genetic engineering accomplishment of mankind that had never been seen or done before. He had opened the door for a new approach to make genes and proteins work for us at our own level of development. Venter's bacterium could be made to work for us, manufacturing chemicals for our brain's dysfunctional organs. In Parkinson's disease, Venter's bacterium could be made to build dopamine, replacing the substantia nigra dopamine deficit. I can see it being adopted in agriculture, where it could be engineered to kill bugs, replacing harmful pesticides. Bacterium engineering could be used to increase food production to feed our world population. We have many variant genes that seem to be idle or not performing at our own level of self-development. Venter has proven that we have grown up; he has proven our humanity and love for one another. We can take charge of our own organism. He just finished sequencing his own genome extremely recently in the time line of human life.

C. Venter Challenged HGP

C, Venter challenged the federal government–sponsored Human Genome Project and came out a big winner. His success relied upon improving existing technology to assemble and improve on nature's already existing life structure. Life itself is dependent on small elements coming together and functioning as a separate unit. These units of life are around us, and the universe is full of it. C. Venter's concept of throwing little pieces of life elements into the cell (genetic engineering) was around for almost four decades. I refer here to recombinant DNA pioneers Herb Boyer and Stanley

Cohen's work. Without hesitation, I can call them the first and most accomplished genetic engineers. The beauty of C. Venter's synthetic bacterium is its ability to engineer organic material from the universe (which our planet is part of) to assist and improve human health and happiness.

When we first cut and isolated a DNA molecule, making use of restriction enzymes, we continued our game with our alphabet of life, throwing in the ligase enzyme. It glued together, and a plasmid came out to exist as another tool. These two new tools of research and therapy, restriction enzymes and recombinant DNA, took over the front pages of scientific journals, newspapers, and newscasts on TV. From building and engineering a bacterium, we can go on and rewire naturally existing endogenous gene circuits and pathways. I am aware that this would be a difficult but gigantic step forward. However, Zhen Xie, a Harvard University professor, along with others, reported a discovery that appears to be worthy of further study and analysis. His team has developed tools we can learn to use at present time. Incidentally, George Church's bacterium engineering might come in handy in producing killer cells without harming normal and healthy cells. Cancer cells are terrorist cells. We should be channeling funds to identify and destroy these terrorists producing cancerous cells. C. Venter, G. Church, and Zhen Xie, along with other pioneers in this field area, are ahead in this race of technological revolution. We need to transfer and apply this technology into practical use for anyone who needs it. We have the technology and the will to share it with others. Funding is in dire need.

When engineers were cloning DNA sequences, cutting with restriction enzymes, and linking them together with a DNA enzyme named ligase, they were making little tweaks to the genome cells. They were using the bacterium cell as the workhorse as it divided over and over again in a culture. Boyer and Cohen were using an existing microorganism for the benefit of a multicellular organism.

We can argue that Boyer, Cohen, Venter, and other scientists are just accelerating and improving on the physiology of eukaryotic organisms. Life on Earth has gone through multiple twists and turns that are not fully clear to most of us. One of those twists must have been when the RNA molecule made advances to another powerful and interesting molecule—the deoxyribonucleic acid molecule, DNA.

Francis Crick and RNA

The brilliant scientist Francis Crick, according J. D. Watson, suggested as early as 1968 that RNA must have been the first genetic molecule and that RNA, "besides acting as a template, might also act as an enzyme catalyzing its own self replication."[9] This insight of Crick on RNA when the double helix and its nucleotides were still trying to find their rightful place in our brains, was awesome. This fellow seemed to pull his ideas out of his jacket's rear pockets, and most of his ideas were proven correct. Even on his way to the hospital before his death, he was doing science in the ambulance, according to some of his biographers. I am grateful we have guys like him working for humanity.

A team of researchers from one of the most prestigious universities in the United States wrote, "The reality of the "R.N.A. World", an R.N.A. dominated stage in the early evolution of life prior to the evolution of coded protein synthesis, has been firmly established by recent structural studies of ribosomes.[10] I must point out that in the protein assembly factory, a ribosome, one of its components is ribonucleic acid. This acid in its multiple molecules alone is not only an effective workhorse. In addition, in the ribosome, it is part of the

---

[9] Raymond F. Gesteland, *The RNA World*, 3rd edition (New York: Cold Spring Harbor Laboratory Press, 2006), xxiii.
[10] Ibid., 57.

machine that builds very complex organisms, such as human beings. I will continue to address the chemical composition of a ribosome. Further elucidating cellular and life complexity on planet Earth, I must attempt to explain how the mitochondrion—the cellular energy engine—came to be part of the human organism's cells. Nature has been working with and engineering our basic functional units, our cells, from the very beginning of life itself. All our basic elements, beginning with carbon, nitrogen, hydrogen, oxygen, potassium, sodium, etc., are part of nature. Watson, Crick, Cajal, Collins, Cohen, Boyer, Venter, and all of us, are nature in its natural and beautiful development. The contributions of pioneer scientists sequencing the human genome has helped our relationship with other species. It has helped us understand ourselves by understanding the genome of our closest hominid relative, the chimpanzee. The mouse, not too far behind, is among the best research tools we have. By engineering and observing human diseases in a mouse, we can trace genes responsible for the disease and develop corrective therapeutic tools.

Current worldwide issues—climate change, for instance—may also benefit from hominid genome sequencing. Anthropologists, historians, and scientists of multiple orientations must have enjoyed Svante Paabo's work at the Max Planck Institute in Germany. His latest discovery is on *Homo sapiens* traveling around our planet. He brought up a new issue with the out-of-Africa theory as part of mankind history on our planet. He posed serious questions about climate changes and their impacts on plant and animal survival on Earth. Neanderthal once occupied most of Europe from the Iberian Peninsula to southern Siberia in Russia. Climate change brought drastic weather changes, and most of the continent was covered by heavy snow all year around. This study relates to us in terms of demographic movement around the world and its impact on us. The sequence of human genome from different parts of the world's population helps us understand how genetics and environments relate to and influence each other in disease etiology. Through worldwide genome studies, we can identify the most common variants from samples in China, Japan, India, Russia, Europe, and the United States. For example, does autism appear in the same gene and chromosome in children in any of those countries? Biomedical technology developed in the United States. The use of multiple RNA tools can bring the East and West together in fighting diseases and plagues in animals and plants. It seems the United States trades with every country on this planet. We eat their foods and handle many of their products, which may allow disease to spread.

After the HGP Completion in 2003

Following the completion of the Human Genome Project in 2003, many surprises linked to genetic structure in the chromosome have come to light. Large segments of noncoding genes are present far away from genes associated with a disease; however, it looks as though they are linked to a disease. In some cases, gene variants common to autism have been linked to brain diseases like schizophrenia and manic depressive illness, also known as bipolar disorder. Autism, like schizophrenia, is a multiple-symptom brain disorder. Multiple genes and environmental causal factors may be involved in triggering its behavior. A group of international researchers studied nearly a thousand autistic persons and compared it with an equal number of healthy individuals. They found dozens of genes involved in autistic disorder. Interestingly, most of the variants in autism were missing segments or duplicated segments of DNA. Some of the genetic changes are inherited, while others have to be attributed to parents or the child himself.[11]

Common symptoms of autistic children are their severe social

---

[11] *Science News* (October 23, 2010) 18–21.

limitations and difficulties, language problems, and repetitive behaviors. "The gene dubbed *MET* regulates production of a protein that influences cell proliferation in various parts of the body. This gene lies on a stretch of chromosome 7 that other researchers had linked to autism."11 Neurologist Pat Levitt from Vanderbilt University, a team leader, consulted databases from 204 families with one or more children with autism. Another group of 539 families was included in the study and found that the link between the Met variant and autism appeared primarily in families with two or more affected children. Laboratory tests showed that this MET form lowers the gene's activity and reduces its production of proteins that binds to various tissues. Daniel H. Geschwind, from the University of California at Los Angeles, praised this report as the first time someone had identified a candidate gene for autism; however, he cautioned that MET could be the tip of the genetic iceberg.[12] A few years have passed since Pat Levitt conducted and published this report. More recent research done by scientists attempting to untangle the genetic roots of autism has found dozens of genes involved in this brain disease. But I emphasize once more that there is more than just genes implicated in this disease. Autism, like schizophrenia, is not limited to race, nationality, ethnicity, or social status. Common efforts to find a cure might not be too far away.

Autism affects mostly boys. Finding out the etiology of a genetic disease does not just mean pointing out gene duplication; missing segments of the genome sequence; molecular intrusion, such as short-nucleotide polymorphism (SNP); or a mountain of noncoding genes. Looking for rare variations in the human genome is only one approach to multiple rare syndromes like schizophrenia and autism. What caused genome segments to be missing or duplicated in the first place? What provokes SNPs to appear in harmful locations? Researchers have shown that not everything is inherited. How

---

[12] *Science News, Vol.* 170 (October 21, 2006): 259.

about the embryonic phase of gestation? The parental mental and physical state at the time of conception undoubtedly contributes to embryonic, fetus, childhood, and adults problems. The human embryo is subject to multiple risks and threats from inside as well as outside the mother's womb.

Such issues are caused not only by gene mutation, such as a letter A being replaced by G. One has to consider epigenetics, a set of chemical alterations not found in the DNA code that partially manipulates gene expression. Among epigenetic markers, we have methylation (CH3), acetylation, phosphorylation, and similar markers that may attach themselves to tails on histones and alter critical transcription processes. Histones are proteins that DNA coils around. Further complicating our job, we have micro RNA that may bind itself to messenger RNA and suppress the synthesis of a particular protein. Suppose the protein was not synthesized properly and remained incomplete or modified in some way and could not take its rightful place in a visual pathway in the occipital lobe. This would definitely put one's vision in jeopardy. In this scenario, outside environmental threats and risks must be considered as secondary to biological causal factors. A biological culprit can be found in any vital organ of one's body; there is not necessarily a dysfunctional gene. Revisiting Allis's hypothesis on histones, DiTacchio and others[13] showed us that the circadian clock that governs diurnal rhythms of physiology and behavior uses Allis's histone code extensively. However, DiTacchio also showed that histone demethylase is important for the activity of the circadian activator complex, CLOCk-BMAL, and for circadian oscillation in general. In addition to Allis's hypothesis, modified histones serve as platforms for adenosine-5'-triphosphate (ATP)-dependent motor complexes that recognize chromatin.

---

[13] *Science* (September 30, 2011): 1833.

DNA Packing

DNA packing could become an area of extreme care susceptible to harmful epigenetic intervention. Histones are proteins responsible for the first level of packing in eukaryotic chromatin. The complex comprising DNA and proteins (histones) that forms a chromosome is called chromatin. The amount of histones in chromatin is approximately equal to the amount of DNA. Histones have a high proportion of positively charged amino acids (lysine, arginine) and bind tightly to the negatively charged DNA, thus forming the tightly packed chromatin. This tightly packed complex is a safeguard against outside intruders. It can be seen as nature's attempt to keep DNA uncontaminated and as complete as possible. Seen under an electron microscope, unfolded chromatin appears like beads of a string or necklace. For those with a religious orientation, it looks like the beads of a rosary. Each bead is called a nucleosome. As you may have already suspected, chromatin organization in a nucleosome may influence gene expression by limiting the access of transcription proteins to DNA.[14]

In summary, chromatin is the molecular substance in all eukaryote cells that facilitates DNA packing efficiency. It protects itself from unnecessarily unwinding itself and exposing its nucleic acids—adenine, thymine, guanine, and cytosine—to risky alterations. The DNA coils itself around histones, which are proteins that look like tiny beads. They form strong bonds with each other through negative and positive charges. However, for reasons we do not understand at the present time, genome development through millions of years allowed histones to grow tiny tails that stick out of their bead-like structure. These tails, like any external protuberance, may serve the molecule well, but the tails can be a tempting target

---

[14] Neil A. Campbell, *Biology* (San Francisco, CA: Benjamin Cummings Publishing Co., 1987) p. 372–3.

for SNPs and other intruders to anchor to, provoking havoc for our life-creating acid, DNA. On the other hand, histone tails may serve as anchor sites for extracellular transport of nutrients for both DNA and histones. It may also serve as a listening post for communication among cells in bodily organs and between neurons in the brain. How the cell keeps a balance between growth factors and SNPs while conserving continuity among the core genes that will maintain the species genotype and phenotype with minor changes is another question researchers have not been able to answer in a satisfactory fashion.

A Seeming Contradiction

It seems a contradiction for us that histones' proteins, around which DNA is tightly wrapped, develop tiny tails that may threaten their own structure and function. Cells are chemical factories going through multiple processes to maintain intrinsic functions. The cell has many channels in its membrane, including the energy-producing dynamic molecule ATP (adenosine triphosphate), which plays a very significant role in metabolic processes. Does it have a role in protecting the chromosome complex—in particular, histones' tails—from undesirable intruders?

Multiple proteins are involved in forming the chromatin complex, including the critical transcription and translation processes that are critically important for protein synthesis. I mentioned previously that multiple mistakes of different sorts take place during these two stages of protein synthesis. Proteins are crucial for DNA tightly coiling around and forming a shape, besides being part of ribosomes. A group of researchers interested, among other things, in DNA and histone packaging in chromatin, studied an enzyme dubbed *Isw2*. It is an ATP-dependent chromatin remodeling protein that is highly evolutionarily conservative. They found that the positioning of thousands of nucleosomes adjacent to important regulatory sequences is controlled by 1sw2. This enzyme is able to use energy from ATP hydrolysis to override the inherent nucleosome-positioning signal of the underlying DNA. The 1sw2 enzyme may function generally to reposition nucleosomes in unfavorable DNA sequences. The ability of proteins such as 1sw2 to reposition nucleosomes provides a clear illustration that cellular factors operate to disrupt the intrinsic cues that would otherwise package the genome.[15]

I hope I have given you some idea, although incomplete, about multiple problems in just trying to identify possible loci for a disease. As mentioned earlier, this involves not just the genes of one parent or the other, but epigenetic factors and everything around us. Continuing with epigenetics, Frances Champagne, a neuroscientist at Columbia University in New York City, has said that DNA methylation is the most enduring of epigenetic modifications. It seems to be a complicated system of DNA synthesis, packing, replication, and transcription. In a base pair, guanine-cytosine, the methyl molecule, CH_3, may place itself next to a cytosine nucleotide in the double helix and become methylated. The problem is that methylated cytosine looks like another nucleotide called thymine and may be converted from cytosine to thymine. Look at this molecular scenario and think about the possible problems in protein synthesis in this DNA mismatch.

Suppose this protein's destiny was either the heart or the occipital lobe, the region in the back of the human brain that controls vision. In all probability, it would provoke serious problems to our eyesight. Another area of great concern for us is the prefrontal cortex—the last brain region to attain maturity. It is vitally important in cognition, rational thinking, and decision making. Recently, researchers have found that the prefrontal cortex is intrinsically connected with the limbic system in decision making. A rational decision is dependent

---

[15] *Nature, Vol.* 450, no. 7172 (December 13, 2007): 1031–5.

on input from both brain regions. During adolescence, the time when we are neither an adult nor a child, hormone release is triggered by simple stimuli—emotions that arise from the limbic system. The emotional stress and physiological demands we go through during this phase of our life development and growth is, perhaps, the most critical of our entire lifetime. Stress may be provoked by many things inside and outside us. Unconsciously, we may be under continuous stress caused by painful childhood experiences; this may make a person predisposed to developing multiple illnesses later in life. Stress may come from parental disputes; illness; natural disasters, such as earthquakes, tsunamis, and fires; poor nutrition; or civil strife. A poor diet may contribute to inadequate consumption of foods rich in amino acids. Amino acids are the building blocks for proteins, our bodies' building blocks.

DNA Methylation

Before I proceed to elaborate on another theme, I would like to say a few more things about methylation. DNA methylation is a modification that controls gene expression. This is a very important and strong statement. If it controls gene expression, consequently, the whole process of protein synthesis is under its influence. Methyl groups, which are molecules composed of one carbon atom and three hydrogen atoms, seem to be powerful molecules. DNA methylation contributes to mammalian development, aging, and some diseases. In humans and in mice, the addition of a methyl group to cytosine within a cytosine-guanine dinucleotide is catalyzed by DNA methyltransferase and the enzymes DNMT3A, DNMT3B, or DNMT1. The latter is a maintenance methylase, because it adds a methyl group primarily to double-strand DNA that is already methylated on one strand. The cytosine-guanine dinucleotide is represented by the letters T and G that I referred to when discussing the double helix.

When a methyl group is added to DNA bases, it will modify the sequences. For example, let us look at GAAT. The methyl molecule CH_3, when attached to the letters just mentioned, will make the letter group (gene) look different from surrounding similar groups of letters. Consequently, restriction enzymes that make cuts when they recognize particular sequences will bypass the sequence GAAT. This sequence of letters can be said to be under the control of methylation. Methyl is not the only DNA marker, and these markers play multiple roles.

You must keep in mind that you have brave and extremely smart cells in your body. Your DNA molecules play around with four letters, or bases; amino acids; and RNA and go through multiple molecular transformations before a protein comes out to form part of your eyes, heart, brain, or legs. These are very complicated processes that are taking place inside you.

Chronic Stress and Your Brain

The prefrontal cortex is often associated with schizophrenia, bipolar disorder, anxiety, and a host of other brain disorders. It seems the cell has insurmountable problems to conquer to be able to survive. Of course, it has developed its own repair system, including demethylation. J. David Sweatt, at the University of Alabama at Birmingham, wrote while doing research on methylation that demethylation occurs rapidly under certain conditions, such as when people experience stress.[16] Chronic stress has been demonstrated to be extremely harmful to our mental and physical health. Brain organs like the amygdalae on both temporal lobes, the seat of learned fear, are groups of brain cells that are prominently exposed to chronic stress.

The hippocampi on the right and left hemispheres of the human

---

[16] *Science News*, Vol. 173, no.17 (May 24, 2008): 19.

brain are sites par excellence for memory and learning that can also be seriously compromised by stress. Some researchers have found that chronic stress kills cells in both hippocampi, thus seriously compromising one's ability for learning and memory storage and retrieval. An overstressed amygdala is not be able to respond properly or quickly when one is faced by a threat. In the jungle, the amygdala saves humans from becoming lunch or dinner for a hungry tiger or lion, or from being bitten by snakes. In big urban areas, the amygdala helps save us from human predators. Similarly, when your defense system is weakened by chronic stress, viruses and bacteria that are otherwise dormant may seriously threaten your health.

Most of us are aware of stressful situations and their harmful effects on blood pressure and heart health, as well as their ability to cause tension headaches. We usually run to the pharmacy to get over-the-counter medication to get rid of pain and discomfort from chronic stress, but we rarely stop to think about the harm we are doing to our bodies. When you are under stress, particularly chronic stress, your entire immune system is seriously compromised, and symptoms and diseases otherwise dormant under the watchful eye of T and B cells and an army of assistant cells can occur. Bacterial and viral diseases do not miss a chance to attack healthy cells and get people ill.

One more note on the amygdala. In recent years, a surge of new research has expanded scientists' views of the amygdala's importance. It turns out the amygdala helps shape behavior in response to all sorts of stimuli, both negative and positive. It plays a role not only in aversion to fright but also in pursuit of pleasure. It has neural connections with all five senses. This is a very primitive organ of your brain that assigns values to rewards and adjusts those values as circumstances changes. Research also suggests that both amygdalae play a role in goal-oriented behavior. Having connections with all senses, they are in a unique position to prompt immediate precautionary action when it is absolutely necessary. In addition, they involve our commander-in-chief region of our brain, the prefrontal cortex, for rational decisions. I must add that connections between the amygdalae and the thalamus are very, very close. If necessary, the axis composed of the thalamus, hypothalamus, pituitary gland, and adrenal gland will pump out hormones all over the body. People are not aware this is occurring when it happens, but the heritable genes they got from their ancestors equipped them with emergency tools for protection against unforeseeable threats. You can rest assured that methylation is implicated during these processes.

Body Energy and Wisdom

Your body and my body have in storage vast amounts of energy and a treasure of accumulated knowledge and wisdom. Have you ever thought of the amount of energy stored in each atom of your body? Ever since we dropped the first atomic bomb in Japan, I have been wondering about the potential power of atoms in my body. Einstein, Bohr, Oppenheimer, and the likes have put ideas in my brain that make me wonder how I exist on planet Earth. I am made of atoms that form molecules that will become cells and ultimately the organs and tissues that form my body. However, my body is more than 99 percent empty, with electrons traveling in orbital fashion at nearly light speed. The subatomic particles traveling at such speed inside my atoms inside my body seem to me to be part of a powerful walking energy warehouse. I have very little control of what is going on in this chemical and atomic machine of mine.

But how can I say that this machine is mine? My organism is continually processing external perceptions and internal chemical and atomic reactions. When I am witness to a car accident on the highway and I see people injured—especially if the injured are children and women—I will have millions of chemical reactions taking place inside me that I am not aware of. Sadness and fear might be the first feelings registered in my brain. The thought that I could have been involved in the accident places my amygdalae in a state of alert.

There is an amygdala in each temporal lobe. The amygdalae will automatically alert my sympathetic nervous system for an immediate reaction if it is called for. If it is life threatening, the amygdalae will provoke a fight-or-flight response if they do not chicken out and freeze. In the meantime, the amygdalae will connect with the thalamus, which in turn will get in touch with the hypothalamus, which will release a message to the pituitary gland, which will release a hormone to the adrenal gland. The adrenaline coming from this gland will move through my body and trigger myriad responses through millions of chemical and electric reactions. Needless to say, my limbic system and prefrontal cortex are involved in all decisions in my brain—except, perhaps, the split-second reaction of the amygdalae in a life-threatening situation.

The point that I am trying to convey through this hypothetical illustration is that a visual stimulus with no matter attached to it is capable of provoking in my organism (maybe yours too) a series of behavioral responses in which matter and electricity are involved. Matter is converted into energy; and energy, such as that in neurons, is converted into matter. I do not know an iota about how this transformation takes place, but rest assured it is taking place every fraction of a second in my body. This stressful situation I just described for you compromises my health in various ways, including the suppression of my immune system. My heart rate goes up, my blood pressure increases, I exhibit asthma symptoms, and I suffer a headache; the stress becomes an unavoidable health threat.

You may dismiss this health lesson as a trivial everyday episode in modern life. At issue is your own reaction to daily life episodes. How many stressors do we encounter in a day, a week, a month, or a year? In this biochemical scenario, your DNA, RNA, and their many component molecules are modifying the structures and functions of cells, organs, and tissues in your body. A hyperactive amygdala that is under chronic stress can be a serious health problem. An excess or deficit of a neurotransmitter in the brain can lead to a brain disease. The sooner we become aware of genes, epigenetics, and environmental causal factors, the better our health and quality of life.

Through worldwide network studies, researchers are sharing genome-connected information regarding diseases. This allows a researcher or research teams to follow up on another person's findings and ideas. Not all researchers have access to money and technology to go to the next level of investigation in their project. Some countries and institutions provide research funds only when a result is close at hand and has an immediate and necessary use, such as treatment of disease or increased food production. Long-term basic research, if not ignored, is pushed aside. In the United States and Europe, basic research takes in a good sum of money. The sequencing of the human genome, restriction enzymes, recombinant DNA, gene therapy, and deep-brain stimulation, for example, are all American accomplishments. James D. Watson put the double helix together in England, but he is an American by birth. Recently our scientists have been exploring how to engineer genes to make proteins, and the areas in which these genes should be located. A new research tool dubbed *ontogeny* would allow neuroscientists not just to observe RNA molecules responding to DNA messages but also to empower DNA to manipulate neurons at will. Gero Miesenböck, at Oxford University in England, says light-responsive molecules used in ontogenetic experiments use a specific wavelength of light. A channel opens and allows positively charged ions to flow into the cell. K Deisseroth from Stanford University said this happens to be the neural code for "on." Other light-responsive molecules, when exposed to the corresponding wavelength of light, allow negatively charged ions into the cell. "Using a combination of the two types of molecules and different wavelength of light, researchers can flip neurons on and off at will to find out how neurons interact with their neighbors.[17]

I hope this new research tool will allow us to look at neurons in the hippocampi and actually observe the neuronal behavior in learning and memory formation. Similarly, focusing on the amygdalae, it will show you learned fear. The nucleus accumbens and the tegmental ventral area in and around the midbrain will teach us how addiction behavior is interconnected in that area of the brain. Researchers are moving forward at a great rate after the discovery of the double helix and the genetic code. They are in pursuit of our goal, deciphering and making corrections on gene misspelling, mutations, and multiple mistakes in translation processes.

Now I will return to SNPs and missing nucleic acid letters—such as A, T, C, and G—in DNA, as this comes up very often in research literature. It is well known that tagging—binding—a SNP such as a methyl group to DNA generally suppresses genes next to it. I already mentioned the possible harmful effects of missing or altered genes in protein synthesis. Rare mutations—single-letter DNA changes—can contribute in a big way, whether a person gets a disease or remains relatively healthy. Some variants have been connected to common disorders like schizophrenia, bipolar disorder, and mental retardation. Determining the genetic base for any of these diseases will take time and perseverance. It is not only a matter of genetics, epigenetics, and nurture; it involves the conformation of the disease itself. In reference to schizophrenia, one may rightfully argue that the whole brain is involved in producing multiple symptoms. The symptoms involve cognitive, affective, visual, and auditory processes and also affect smell, taste, and touch, among other things.

---

[17] *Science News, Vol.* 177, no. 3 (January 30, 2010): 18–21.

This makes it hard to attribute schizophrenia to a single word of misspelling in the DNA alphabet. However, general wide studies with many participating researchers engaging thousands of patients worldwide might point in the right direction while a particular disease symptom pathway is being pursued.

I love to repeat teaching lessons from James D. Watson. He has never stopped working for us. His goal was not to leave for us the double helix but to sequence our genetic code as it relates to our health. He wrote that a single base change in the DNA sequence of a human beta hemoglobin gene results in the incorporation of the amino acid valine rather than glutamic acid into the protein. This simple difference causes sickle cell disease, in which red blood cells become distorted into a characteristic sickle shape.[18] Watson was daydreaming through space and time during his work at Cavendish Laboratory along with F. Crick in the early 1950s. I remember using a sickle-shaped blade to clear weeds and small bushes growing under coffee trees when I was growing up. In some Latin American countries, the economy is based on sugar cane, tobacco, and coffee crops. Therefore, the sickle-shaped blade was a valuable tool.

The Inquisitive Brain

Once more I will repeat this mutation process. We need three nucleotides, or letters of DNA the alphabet, which normally incorporates four letters—A, T, G, and C—for binding to each other. Three letters form a codon. In most cases the genetic code for each amino acid can be multiple codons composed of three letters. For example, the amino acid leucine can be formed by six different codons, while tryptophan is formed by only one codon, namely UGG. Scientist Vernon Ingram at the Cavendish Laboratory had found out that the amino acid glutamine, formed by codons

---

[18] Watson, *DNA*, 67.

GAA or GAG found at position six in the normal protein chain, was replaced by valine, which can occur in four different codons of the genetic code—GUC, GUU, GUA, and GUG—in sickle cell hemoglobin. You must remember the four letters found in DNA and DNA's difference from the RNA chain. RNA has uracil as a nucleotide, while in DNA, thymine is a complementary nucleotide to adenine. In RNA, it is uracil that complements adenine. These two great molecules work together to build tissues in the body. DNA polymerase unzips DNA during the replication process; actually, it is specifically DNA helicase that unzips it. However, during the transcription process in the formation of a messenger RNA, the RNA polymerase helps create a single strand. Messenger RNA carries the message and is the template for protein synthesis.

This messenger RNA that was transcribed from the DNA cell nucleus will end up at a ribosome. The sequence of the messenger RNA will be used at each ribosome to form a new protein molecule. Another molecule, transfer RNA, which carries amino acids, will cross the cytoplasm and anchor itself at ribosomes. At one end of this tRNA is a set of letters—let us say UUC—that recognizes and establishes a bond with its opposite codon, AAG, in the messenger RNA, with the resulting codification for lysine. An RNA molecule serves as a template for translation of genes into proteins. Another RNA molecule transfers amino acids to ribosomes, the assembly sites for protein synthesis. It also conforms into other useful small forms. Additionally, as previously mentioned, multiple codons can specify for the same amino acid. This dynamic process could create problems in protein synthesis. Take, for instance, leucine, which has six different codons, and tryptohan, which has only one codon. It would appear that protein folding from tryptophan would be much less prone to mistakes than leucine. Now consider variants like nucleotide polymer components, their availability in the cell, and the intervening stages in codon selection, availability, peptide chains, and final protein folding; there is plenty of room for multiple genome mistakes and consequent diseases.

You may want to accuse me of picking on translators, but this RNA transfer molecule has to do a superior job not to commit mistakes culminating in dysfunctional proteins and diseases. Improperly folded codons, competing codons, interference RNA, and microRNA, just to name few of the possible roadblocks in our endeavor, are presenting possible risks in mammal genome sequence studies. I hope I have given you a good idea of how nucleic acids form proteins for your body while you are oblivious of the complex processes taking place. A couple of pages back, I mentioned that this transportation system, seemingly so simple, can be threatened by small nucleotide polymorphisms (SNPs) and interference RNA, preventing genes from carrying out their job of protein synthesis. These SNPs molecules love to mess around with the thymine nucleotide in the double helix and create problems for us.

Epigenetics

EPIGENETICS, THE STUDY OF the nucleotide process that may control a gene's potential to express itself, has become a cover story in some journals because it has been implicated in some common but hard-to-break diseases. The most common and best studied form of epigenetic intervention in a DNA molecule is methylation. It has the potential to manipulate the cell to ignore any gene in a stretch of DNA. We are subject to epigenetic interference from very early in our life processes, according to some researchers. It can begin during our embryonic phase of development. During this early stage, when our cells are undifferentiated, stem cells accumulate, among other things, methyl groups, which direct stem cells into one of the three germ layers. Each of these germ layers produces a different set of adult tissues. These tissues may be protein for your heart, lungs, eyes, or brain. Making epigenetic a little more interesting, but headache provoking, a group of researchers at the University of Sydney in Australia discovered that methylation of the fur color genes in mice persists in the female germ line, allowing the color to be passed down to offspring like a change in DNA.[19]

Local and worldwide studies share their common interest in the human genome, epigenetics, and, particularly, methylation interference in DNA expression. A lesson to be seriously considered

---

[19] *Scientific American*, Vol. 303, no. 2 (August 20, 2010): 2224.

during disease etiology research is the role SNPs in all forms play in methylation, interference RNA, and microRNA during embryonic and fetus development.

The cell has proven to be an excellent self-sustaining unit of life, but there are many internal and external factors that threaten good health. Whether it is endogenous or exogenous, the great DNA molecule has survived millions of years. It developed brain cells for intelligence and critical thinking. This brain is developing very sophisticated tools for self-correction and for the acceleration of its own growth. Among our recent achievements in improving our health is the decoding of the human and mouse genome a decade ago. Mice and humans have similar organs, making the mouse a first-class research tool to study illnesses that afflict humans. Most scientists agree that a mouse genome sequence is about 95 percent the same as a human's. We cannot play around with a human as a scientific research object, but a mouse can be carried with you even when you go on vacation. With gene-targeting technology, we can knock out a gene in any organ of the mouse and peacefully and intelligently observe any causal effects without any pain being inflicted upon a human being. Besides, a mouse does not require lodging, dining, or recreational facilities. The researcher and his or her tools are the expensive part of such experiments.

These knockout mice are very important in medical research. Using a mouse as an object for gene studies is relatively new. The first report about this technology being able to generate gene-targeted mice was published in 1989. Twenty-one years later, three pioneer scientists working independently of each other were awarded the Noble Prize in Physiology or Medicine. The three pioneer scientists are Martin Evans, from Cardiff University, Wales; Mario Capecchi, from the University of Utah, and Oliver Smithies from the University of North Carolina.

Sequencing genomes from different species can tell us how each one of us is different from each other's genotype and phenotype. It can point out mutations and diseases. But perhaps most importantly, comparing genomes from various animals can identify genes and gene segments that have remained unchanged for many years—meaning perhaps millions of years. We can date changes in animals based on a set of genes that have remained unaltered over millions of years. By observing gene locations for protein expression in various animals, we may be able to tell how a species changed into a different phenotype. However, the immediate benefit of comparing genome sequences is the increased precision with which researchers can reveal the sequences that have been carefully preserved over time, implying that they have an important role in the organism. Alternatively, these comparisons can pinpoint sequences that differ in just one species or a group of species.

More on Epigenetics

To some of you, I may sound unnecessarily repetitious and perhaps cognitively naive. My apologies to you, but my classroom experience and supervision taught me that going over a subject or a lesson is not a waste of time, as it strengthens synapses. I will define epigenetics as the study of heritable changes in gene expression that are not the product of changes in DNA sequences. I briefly addressed epigenetics indirectly when I mentioned SNPs and gene modifications. I also spoke about the chromatin complex composed of DNA and histones with its little tails extending out as possible SNP sites. I attempted to connect it with current concept of protein switches. I could easily compare this concept to switches commonly used in electrical systems; it would sound and look neat, clean, and easy to follow. However, we are dealing with living acids like DNA and RNA, which are capable of forming multiple tissues and organs.

During these processes, including before and after transcription, many modifications take place, altering the ultimate product—proteins. Where and how these modifications take place has been a subject of curiosity among researchers during the last decade. Small-nucleotide molecules attach to chromatin most often, linked to or hanging on to histone proteins; these are called epigenetic marks by most researchers. Many unanswered questions are still lingering around the concept of the enzyme switches. Are epigenetic modifications passed down to offspring through cell division? What about modifications that take place during the transcription phase, which the two great acids DNA and RNA are involved in?

Before I proceed to create more questions, let me deal with chromatin modifications. One of the concepts most frequently used to explain protein switches is that of DNA and histones modifications being consonant with either negative or positive transcription states. In histones, posttranscription modifications (a cytoplasmic response to transcription factors—positive marks) are established during the gene activation. It takes place during gene activation by recruiting relevant enzymes by DNA-bound activators and RNA polymerase. Similarly, negative marks are established across genes during repression by DNA-bound repressor recruitment.

This seems to be a simple and attractive model that could be used in conducting chromatin research. However, some researchers claim that it is not as simple as it appears to be. The concept of enzyme switches is not a matter of connecting and disconnecting cables or just turning on genes. Many of the marks I have been talking about play multiple roles. It is a dynamic metabolic process in intracellular chemical reactions that converts an organism's infinite number of molecules into energy-producing action units. Chromatin marks are more complex molecules than anticipated. They respond to and performs multiple complex roles. It is now a belief among some researchers that chromatin's role in transcription is that of a mechanism in which DNA and histone modifications establish gene activation and then reinstate repression. Thus, DNA methylation and histone modification is a very dynamic process; it precludes a fixed status. Some marks now seem to recruit both, activating and repressing effector proteins. Besides, positive and negative acting marks are established during transcription. Two types of chromatin modifications for the regulation of transcription of the protein-encoding genome are activation and repression. These DNA modifications seem to take place at specific junctions of nucleotides and at different locations.

Posttranslation Modifications

There are a good number of posttranslation modifications. Most take place in the amino and carboxyl terminal histone's tail. Among common modification sites, lysine is a key substrate residue in histone biochemistry. It undergoes many exclusive modifications. This really seems to be the Achilles tendon I referred to at the beginning of this essay. "In general, the functional consequence of histone's post-translational modifications can be direct, provoking structural changes to the chromatin. Changes can take place by recruiting proteins."[20] Epigenetic changes are decisive and are indispensable during the development and differentiation of various types of cells in any organism and normal cellular processes. However, as in all systems, an inappropriate epigenetic modification can lead to serious health problems. Epigenetic processes may be involved in very sensitive areas, such as in cell shape and function. They may intervene in chemical modification to the DNA and histones in the chromatin packaging. This macro-molecule, DNA, has taken appropriate measures to protect itself from harmful intruders. Its job is awesome. The protein that DNA wraps around to form the chromatin structure, histone, has itself gone through multiple

---

[20] *Nature, Vol.* 447, no. 7143 (May 24, 2009): 407–12.

processes. Proteins are the most abundant molecules in our bodies. The number of proteins in our bodies makes them the most likely targets for changes, modifications, and assault by internal and external forces.

The prefix "epi" means "over" or "in addition to." "Epigenetics" is now used to refer to the study of heritable changes in our appearance. In genetics jargon, this appearance is known as "phenotype." Such changes may also appear in gene expression, but they may not involve changes in DNA sequence. At present, the word *heritable* is paramount in epigenetics as long as it does not involve changes intrinsic to the DNA sequence.

You may rightfully wonder how it can be heritable if it does not involve the DNA sequence. Well, epigenetics has become a field of study of biology in itself. Epigenetics, like RNA's multiple forms and functions, has attracted brilliant biology researchers hoping to find and characterize toxic proteins that threaten our health. We have just finished sequencing the human genome; searching for tools to destroy pathogens that threaten our organism is our next challenge. The job ahead is worth our interest and enthusiasm. It took our most primitive cell millions of years to evolve from a simple cell to a complex organism like yours and mine. Among your body's multiple organs, you have a brilliant brain that is engaged in finding solutions to diseases that threaten your life. Your brain, when not engaged in fighting diseases, amuses itself trying to become a painter like Picasso, a violinist, or an astronaut navigating outer space.

Maternal Nurturing and DNA Methylation

My main concern during this essay has been centered on how genome sequencing can improve human health. There have been well-documented studies in mammals reporting that maternal nurturing alters DNA methylation at the gene-encoding glucocorticoid acid receptor. In the absence of appropriate nurturing, the chemistry of DNA in her offspring is seriously compromised or placed in jeopardy. If normal DNA methylation is negatively altered in genes coding for proteins destined for the hippocampi, the child's learning ability will be placed in jeopardy. The same is true for negative DNA methylation for genes coding proteins for the amygdalae. In all probabilities, the child will not be alert enough to appropriately respond to environmental learning signals. Whether it is the hippocampus or amygdala functioning deficiently secondary to abnormal DNA methylation, the child's general health will be at great risk.

Applying the definition of epigenetics as heritable changes not involving the DNA sequence helps us understand and deal with learning difficulties and emotional problems in our schools and general population. A considerable number of research studies have raised the profile of epigenetics, making it a particular field of study in biology. I hope it will help us understand and provide appropriate treatment for diseases like Parkinson's, Alzheimer's, schizophrenia, autism, and violent behavior leading to bloodshed and fatal episodes.

Cancer and Epigenetics

A report from Epi Gen Western Research Group at the University of Western Ontario dated January 21, 2006, defined "epigenetics" as the study of heritable changes in gene expression that occur without a change in the DNA sequence. This definition has been established worldwide as the gold standard for research and findings on epigenetics. The leading author of this report, Dr. David Rodenhiser, states that "epigenetic mechanisms are critical components in the normal development and growth of cells. Epigenetic abnormalities have been found to be causative factors in cancer, genetic disorders and pediatric syndromes as well as contributing factors in autoimmune diseases and aging."

This quotation lists epigenetic abnormalities as a causative factor for cancer. Cancer has been under the microscopic eye for quite a long time. However, there are various types of cancer that appear in various tissues and organs of the body. Men can develop prostate cancer, which tends to stay localized, while many women develop breast cancer, which often metastasizes. There is also colon cancer, which can be removed by surgery. There is throat cancer and lung cancer, both of which are strongly associated with excess smoking. There are cancers whose etiologies are attributed to environmental factors, such as toxic gases, air pollution, water contamination, pesticides, and even chronic stress. However, most of these cancers just named are of a temporal nature or etiology and do not fall under epigenetics because they are not passed on to offspring. The key word "heritable" cannot be applied to a cancer in a person with a long history of smoking a package of cigarettes a day. The same is true of a person exposed to toxic or radioactive chemicals at his place of employment. There is enough evidence to connect lung and throat cancer with cigarette smoking. The leading author in this article proceeds with DNA methylation and histone modifications, describing the basic components of a nucleosome, namely DNA wrapping around clusters of proteins—histones.

Genetic Markers

These coils of DNA around histones are widely known as chromatin. Changes to the chromatin structure influences gene expression—that is, protein synthesis. If protein synthesis is jeopardized, it means serious trouble for the organism. Protein shape determines function, and if the shape of a protein whose final destination is the heart—specifically a valve in the heart—is altered, and if the fetus survives in the mother's womb, he or she will develop heart problems as an adult. I should add that cells have their own intrinsic repair system, but unfortunately the repair system does not work to our advantage every time we are in need.

Let us look at the example of a protein destined to be part of an axon in a brain cell. Axons are engaged in electrogenesis; their functional components are sodium and potassium pumps that will send a chemical message across a synapse to a nearby postsynaptic cell membrane receptor. Any process or epigenetic mark that messes up the formation of proteins threatens the life and health of an organism. The negative effects of missing genes, missing segments of genes, the substitution of one nucleic acid for another and more all provoke health problems.

Further elucidating on genetic marks, Dr. Rodenhiser named some of the enzymes involved in the process, such as DNA transferase, histone deacetylase, histone acetylases, histone methyltransferase, and the methyl binding domain protein (the ending "-ase" denotes an enzyme). Dr. Rodenhiser alerts us that alterations in these epigenetic patterns can deregulate patterns of gene expression, which results in profound and diverse clinical outcomes. Some research findings in David Rodenhiser's article grabbed my attention because they had been confirmed by other research teams.

DNA methylation involves the addition of a methyl group to a nucleotide—for example, the methylation of cytosine by CH_3 within CpG. Those last three letters refer to cytosine–guanine pairs. Cytosine is a nucleotide present in both DNA and RNA. The point I am trying to show you is the importance of this epigenetic mark involvement in critical gene processes.

Another emphasized point I would like you to remember is that changes in DNA methylation may occur as a result of low dietary levels of folate, a B vitamin; methionine, an essential amino acid; or selenium, a lack of which can have profound clinical consequences, such as neural tube defects, cancer, and atherosclerosis. This last observation was supported by five different research groups. Among the factors highly suspected to contribute to altering DNA methylation and histone modifications, adding to the above-mentioned health risks, are air pollution, water contamination, pesticides, chronic stress, excessive exposure to dangerous chemicals, and psychological causal factors.

"Epigenetic mechanisms include DNA methylation and changes in chromantin structure, non-coding RNA, and nuclear organization. The epigenetic mechanism commonly implicated in heritable transmission of a phenotype is DNA methylation". (Science, Vol. 345, august 15,2014, p.733) In the same journal, vol., 348 april 3, 2015, p.90, Raushik et, all wrote: "Changes in histone translational modifications are associated with epigenetic states that define distinct patterns of gene expression.... Previous attempts to separate the inheritance of epigenetic states from sequence specific establishment suggest that specific DNA sequences and DNA binding proteins are continuously required for epigenetic inheritance. In addition, it involves self-reinforcing interaction between histones modifications and RNA interference or DNA methylation". Several epigenetic researches have shown us how Epi, is related to some human diseses. Therefore, I feel approiate to add another quotation. Proteins that change chromatin structure...provide an additional layer of regulation and are considered major epigenetic determinants of cell identity and function". Science, vol.352, june 3, 2016, p. 1188.

The DNA Structures in the Brains of Our Scientists

The idea of the DNA structure and function occupied the minds of a group of brilliant young scientists right after WWII, including our star chemist Linus Pauling. However, the Britons Rosalind Franklin, Maurice Wilkins, and Francis Crick and the American J. D. Watson were determined to give the United Kingdom the prize of stardom in molecular biology. Although Rosalind seems to have been ahead of the group in designing the double helix for the DNA molecule, it was Watson who did not miss out on a single opportunity to check out everyone else's ideas and try to put them to work. He checked out every article in biology and related sciences that might give him the final answer. He alone argued how the spiral strands should be oriented, inside or outside, with the nucleic acids hanging on like necklace beads.

On the other hand, Rosalind may have been closer than anybody else to the DNA double-helix structure, but she insisted on more experiments before announcing it publicly. In Watson's enthusiasm to share his curiosity and observations with Crick, he did not miss an opportunity to pick up cues from other researchers that might advance his own insights. Rosalind's crystallographic pictures struck him like no other pictures had. He finally figured out that it must be a two-strand structure with nucleic acid beads inside that were chemically attracting each other. His approving colleague, F. Crick, gave him his blessing, and the structure of our genome was soon to become a top story in the most prestigious science journals, newspapers, and television networks. I, for one, would have thought that Watson and Crick would be celebrities to be paraded on Broadway in New York City and in Washington, DC. Well, things do not always come to pass the way one would like them to. As if it helps anything, the person behind all these genetics, Gregor J. Mendel, a monk, was not recognized even by his own colleagues in the monastery. In Watson's case, researchers working on and around his discovery were awarded the Nobel Prize for subsequent discoveries relevant to DNA structure and function. Watson and Crick had to wait for another few years before being called by the Nobel awarding committee in 1962.

What was so mysterious about this double-helix structure? The solution of the mystery would unlock the key to the creation of life. Watson and Crick proposed that each nucleotide would chemically attract the opposite complementary nucleotide. Bingo, this was it. Now they needed to proceed with protein synthesis. How could these four letters of the double helix—A, T, C, and G—arrange themselves to make the twenty amino acids necessary in the human body into proteins?

Crick began to work with Sydney Brenner, and together they demonstrated that the code was based on a series of three bases named codons. Researchers Marshall Nirenberg and John Matthaei identified the first letter in the genetic alphabet. They reported that the amino acid phenylalanine had the code UUU. Please turn to the table at the end of the book showing amino acids and their codon codes. You can observe that most amino acids have more than one codon code. In addition to the sixty-one codons for amino acids, there are three codons for a stop command. "The stop codon indicates the end of an amino acid sequence for a given protein. Most proteins comprise one hundred or more amino acids, and so a long sequence of codons is needed for any given protein."[21]

Watson and Crick Parade in the Canyon of Heroes

After discovering all the codons for each amino acid that will form peptide chains, which in turn become proteins for cells and tissues in our body, what was next? A sequence of codons codifies the production of a particular protein. These sequences of codons that code for enzymes regulate neurotransmitters in our brain. Dopamine, a neurotransmitter that is involved in very serious brain diseases like Parkinson's and Alzheimer's, is also engaged in making us feel good. This brain chemical is often associated with feeling

---

[21] Nancy C. Andreasen, *Brave New Brain* (Oxford: Oxford University Press, 2001), 101.

high and addiction. This discovery opened up not only a new field of research—molecular biology—but also another approach to therapeutic medicine. According to Crick's central dogma regarding DNA, RNA, and protein, it is the DNA nucleus that issues orders for protein synthesis in the cell. These proteins are the building blocks of our bodies. No wonder I was expecting a big parade for Watson and Crick through the Canyon of Heroes on Broadway in New York City.

In retrospect, I thought that as soon as we had learned the step of codon sequence and protein synthesis, we could begin engineering cell organelles and construct a miniature life system. I was not too far off; C. Venter, in the year 2010, announced publicly that he had produced a synthetic bacterium. And James Watson was around to celebrate it. Venter demonstrated that we have mastered the mechanism of the DNA and RNA molecules to make live cells that will grow to become whole, beautiful organisms like Dolly, the cloned sheep. Perhaps we will have to wait for Venter's disciples to employ his findings before we see the application of his newly discovered tool for therapeutic medicine and pharmacology. In just a little over a half a century, we have gone from riding mules to flying in supersonic aircraft. When I was attending elementary school, a trip to the moon was thought of as akin to riding on a broomstick; it was a story associated with witchcraft and evil spirits.

The Very Old DNA and RNA Molecules

The DNA molecule began life many, many years ago either by chance or because of a basic chemical and functional need to pair with or incorporate another great molecule, RNA. That union facilitated, among other things, the formation of a protein and growth of a single cell into a multicellular organism. Equally important was the development of an important organelle, the mitochondrion. When and where these wonder molecules fused together is basically an intellectual mental exercise. DNA survives by self-replication, or duplication. During replication, it divides into two strands that serve as a template for transferring the entire database into new cells that will create new organisms. Secondly, DNA functions take place in the cell nucleus. A DNA single-strand messenger, RNA, is formed, which eventually will carry the message for protein synthesis at a ribosome. Another RNA strand, named transfer RNA, travels to a ribosome. Out of this assembly station, the ribosome peptide chains will come out forming proteins. This linear chain of peptides will be shaped into proteins for different organs and tissues in the body. There are many proteins that perform many various functions in various parts of the body.

Introns and Exons

Within the genome, we have coding genes and noncoding genes. The old name for noncoding genes was junk DNA. The new name for a segment of noncoding DNA is "intron." In contrast, exons have the codes that determine the sequences of amino acids that will provide the final protein. This is not an easy and simple job. Just imagine the complexity of a template for a protein destined for the toe of my left foot and a template for a protein in the prefrontal cortex. As already mentioned, most of our genome consists of introns—or, as you may say, junk DNA. Some researchers claim that over 90 percent of our genome is noncoding material.

An interesting mechanism exists within each cell. During the transcription process that produces the mold for protein synthesis, meaning the messenger RNA molecule, introns are thrown out. They are removed before they are transported to a ribosome, where proteins are made. Introns do not have a job in the translation process. They are noncoding genes. The transfer RNA strand carries a three-nucleotide sequence known as an anticodon. It facilitates the matching of mRNA and tRNA at ribosome sites during the making of a protein.

Just stop for a few seconds and think about all the different proteins in the human body. Proteins responsible for making up our brain cells have to, among many other functions, generate electricity to produce impulses that will produce neurotransmitters, which in turn will activate other cells. Your memories from birth to death will be stored in proteins. And how marvelous it is that the instructions or commands from all these proteins, each one with a particular function, come from the nucleus of the cell. This great cell with its powerful nucleus is a very old one and has proven to be immortal. It has been feeding on chemical elements existing on planet Earth for billions of years. The human body is a huge and complex chemical machine that has survived earthquakes, tsunamis, hurricanes, electrical discharges of immense power, multiple ice ages, and a huge meteorite about 68 million years ago. The body works from conception until death. It rests while we sleep, although our vital organs are on a 24-7 schedule, while genes are turned on and off as needed.

"Sleep" is not an accurate term in dealing with cell activity. For instance, take the hibernation cycle. Some small animals hibernate for months or even years, deep in the ground. This is an excellent research area for sleep and insomnia studies. Somatic as well as brain cells show close to zero activity during hibernation, but they are still alive. Buried deep in the ground, they are not responding to outside stimuli such as wind, rain, sunlight, and drought but are maintained by a balance of temperature inside the Earth and protection from whatever is going on above it. Cells engage in a minimum of activity to be able to survive hibernation. Are they sleeping, unconscious, in a comatose state, or hypnotized, or are they just a bundle of atoms clustered in molecules that will begin exhibiting chemical attraction with a little touch of nature? Here we may find the answer to the beginning of life on Earth.

Thalassemia, Hematopoiesis, and Other Related Disorders

GENES MAY BE TURNED on to produce protein to consolidate a memory permanently or to reconstruct a broken ligament in the left toe. Genes may be called in to produce enzymes that will provide chemical reactions that in turn will trigger serotonin release and make one feel sad. All these switches—on-and-off switches—may be provoked by internal or external causative factors.

I once saw a puppy that had been run over by a car trying to move while dragging its paralyzed legs behind it. I caught myself wiping tears from my eyes. My body was responding to the suffering of a little dog I had never seen before. A visual stimulus had triggered a very complex response process I could not control at a conscious level. The cell nucleus was responding to and commanding multiple actions that I was not aware of.

Besides shedding tears for a little dog I had not seen before, I inherited thalassemia traits from my parents. I passed these traits on to my daughter. I was one of eleven children; five inherited it, and one died before he reached his third birthday. In thalassemia, there are not enough red blood cells to carry oxygen. The protein hemoglobin, which transports oxygen from the lungs to the tissues, is made in the bone marrow by stem cells. When my body needs to make more of the protein hemoglobin, the pertinent part of the bone marrow DNA—meaning the hemoglobin gene—opens up, or as the professionals say, unzips. During this unzipping process, only one strand is copied. This process is also known as transcription. Here a protein called RNA polymerase lends a hand in the process, and the single strand coming out of the bone marrow DNA is the single strand the original biologists called messenger RNA. This is the template for the production of a protein. I hope that in the near future, my daughter and family members can benefit from gene therapy and eliminate that health risk from our protein factory.

Protein is put together during the translation process. The command to produce protein ultimately comes from the nucleus. However, what or who is responsible for alerting the bone marrow DNA that there is a need for more hemoglobin is as yet unknown. We do know that proteins turn genes on and off at certain locations. Here I would like to quote from the New York Academy of Science 2005 report on hematopoietic stem cell research: "Despite the intense research, many long standing questions of experimental hematology remains unsolved ... Real time tracking of individual cells in culture, tissues or whole organism would be an extremely powerful approach to fully understand the developmental complexity of hematopoiesis. A vast array of cellular and molecular tools, RNA and protein expression data, and experimental methodologies have been developed for it and are already available for application. However, despite its therapeutic importance and long standing research efforts, hematopoiesis research has been unable to satisfactorily answer many questions."[22]

The Pasteur Institute

I have kept my focus on how genes are turned on and off—more specifically, how genes are called to produce hemoglobin proteins for

---

[22] *New York Academy of Science, Vol.* 1044 (2005): 201–2.

oxygen transportation. I reasoned out that genes respond to internal and external triggers or stimuli to produce proteins. In the process of protein synthesis, you must have an amino acid group at one end of the transfer RNA. I looked for an amino acid in vitamins, minerals, and dietary supplements rich in amino acids. However, my greatest concern was how to turn on the hemoglobin-producing gene. At this level of research, I concluded that the concept of an on and off switch was a very simple concept. The whole process of protein synthesis involving many functional organelles within the cell points to complex system of cell metabolism and chemical intervention.

While I was watching my son Joe work on an electrical switch in the basement of his home, I asked him how he would relate the electrical switch he was working on with a gene switch in our cells. He directed me to the Pasteur Institute in Paris. He added that while he was studying medicine in Spain, a friend of his named Alejandro had gone to Paris for a short trip. Alejandro had good connections and was able to visit the Pasteur Institute. Working on a different type of switch there were a couple of brilliant scientists, Francois Jacob and Jack Monod. The intestinal bacterium *Escherichia coli*, E. coli for short, was receiving their undivided and focused attention. They found a repressor (a protein) that, when bound to E. coli DNA, prevents the enzyme that assembles messenger RNA from doing its job. The process of transcription was interrupted by a molecule at a DNA switch. This was eye opening, if not brain boosting.

Protein shape determines function. I concluded that a DNA child, meaning a protein, could also play a role in regulating gene activity. This was another avenue of study—one that would perhaps lead to an RNA tool capable of engineering or manipulating human genes. Monod and Jacob's discovery made me even more curious, if not restless, about the behavior and function of this RNA molecule. It is playing many undisclosed games right inside our bodies. I was very happy to learn that soon, we may be able to engineer E. coli at will. How E. coli lives in my intestines without paying rent while eating my food, I do not know. E. coli has become an excellent research model worldwide. We provide E. coli food and shelter, which justifies our putting it to work for us.

However, the RNA was interfering with the building blocks of my body. For example, transfer RNA can block protein synthesis, microRNA can do likewise, and much more. From the very beginning, I became suspicious of RNA. Too much power in one person or object gives me goose bumps. But there was one person I could turn to—Francis Crick. Most researchers call him the father of the central dogma regarding DNA, RNA, and protein. Is it DNA that makes use of RNA to do its job, or is it the other way around? How would you answer this question? If you have any doubt, even peptide bond links—one amino acid linked to another—are formed by RNA. The cell nucleus sends out messages and command orders for protein synthesis. However, the fact that there is an RNA molecule that can hijack the messenger strand and prevent the normal formation of a protein in my body is worrisome for me. It was more than I could swallow at the time. Anyhow, there are enzymes—again produced on orders from the cell nucleus—that can influence gene expression. In fact, enzymes play significant roles in a cell's metabolic and chemical reaction. In order to bring to closure the DNA–RNA dispute, I will refer you to a very reliable source that says, "The reality of the RNA World, an RNA dominated stage in the early evolution of life prior to the evolution of coded protein synthesis, has been firmly established by recent studies on the ribosome."[23]

---

[23] R. F. Gesteland, *The RNA World*, 3rd ed. (New York: Cold Spring Harbor Lab. Press, 2006), 57

Jose Morales Dorta, PhD

Harry Noller and RNA

A scientist who could not resist the temptation of further studying the function of RNA, Harry Noller, stripped the protein-assembly station, the ribosome, of all its protein. However, despite lacking the protein that was normally part of its structure, the ribosome was capable of forming peptide bonds. Returning to Jacob and Monod's laboratory and their prized research model, E. coli, they came to the conclusion that DNA must have a control system. Actually, the control system is a protein that activates a gene, which will in turn say, "Let's get to work." Similarly, there must be another protein molecule to stop the gene from continuing to produce more of the same. Thus, the on/off switch was attributed to these two gentlemen who worked with parasites that have made my intestines their home for life.

We know that environmental factors can influence a protein to change its shape to provide a response to an external stimulus, such as a very stressful situation. We also know that stress not only influences our responses through shape changes in proteins but, most importantly, can also kill genes that produce the protein we need in our body. My forty-five years of experience working at a mental health clinic where I applied scientifically proven methodologies qualify me to write on harm done to a person exposed to chronic stress. There are at least three components of the brain that are extremely susceptible to stress: the hippocampus, the amygdala, and the prefrontal cortex. These three organs are considered by most neuroscientists as part of the limbic system. As you already know, the limbic system is known as the home for emotions. It also happens that emotions are a very important attribute of human beings. Through emotions we socialize, make friends, enjoy ourselves at parties, go on vacation to enjoy a couple of weeks away from stress at work, and fall in love.

Another attribute that separates us from the rest of the hominids is language. Through words we communicate our feelings and our emotions, and we have developed signs to write down our emotions. We keep our autobiographical information through episodic memories recorded in our brains, and as a family, tribe, or nation, we pass down our emotions in written form. Emotions, if expressed in excess, can be overwhelming and drive a person nuts. When we fall in love, it seems that the prefrontal cortex, our most rational cluster of neurons, is overruled by the limbic system. Likewise, when anger takes over rational thinking, an individual may commit injury to himself or others.

You do not need me to tell you that a bugging stressor at home, a job, or a neighbor can get your adrenaline to hijack your body and drive you nuts. The same argument can be applied to people predisposed to brain diseases like schizophrenia and depression, heart problems, hypertension, etc. Anxiety does not need a specific trigger to get you upset. We live in a world that provokes anxiety 24-7. We turn on the radio or television, and stress triggers get to us through our ears, eyes, and skin.

I began this journey on DNA and RNA not only because I am a fan of Watson, Crick, Collins, and Venter but also because they have helped me understand myself through their work and throughout their professional careers. It has been my main concern to understand the people I saw during my clinical practice as whole entities. The people I see in my office are made of the same elements I have in my body. They are subject to the same daily threats and risks I face. We may ride the same subway, take the same bus to and from work, breathe in the same polluted air, and eat the same fast food from the same restaurant. We live in cities with the sirens of police cars, ambulances, and fire trucks alerting us to possible life hazards.

Now I come to a personal experience. I come from a family with a history of thalassemia minor. Thalassemia is an inherited genetic disease; there is no cure for this genetic disorder. The disease is classified as thalassemia major and thalassemia minor. I am a carrier of thalassemia minor traits. I inherited the disease from my parent's genes. What concerns me most is that about 50 percent of my family—five out of eleven children—were positive for thalassemia minor. Jack, my brother born just immediately before me, died before he reached three years of age. According to my mother, he died from anemia. My oldest brother, besides being positive for thalassemia minor, died of leukemia after he reached sixty years old. Two of my oldest sisters had children and grandchildren who inherited thalassemia. My oldest sister's granddaughter died of leukemia at age nine. My second oldest sister had her grandchildren follow up at a local health clinic for thalassemia minor.

Both thalassemia and leukemia are blood diseases. Therefore, the suspicion that there may be a connection between these two fatal blood disorders haunted me for some time. Thalassemia is caused by a dysfunctional or a mutated gene. However, in leukemia, a blood cancer, the causal factor or factors for this fatal disease are not fully understood, as is the case with all cancers. Some studies have implicated genes in its etiology. However, there are many different types of cancer, which makes it difficult to link its etiology to any single gene or group of genes. As I said earlier, cancerous cells are rebellious cells that proliferate, invade, and destroy healthy tissues and organs in our bodies. The best tools in the modern medical technology inventory have not been able to identify and explain what originated the initial proliferation of malignant tumor cells. There are many endogenous and exogenous trigger factors that make it very difficult to identify a specific or single causal factor. Does breast cancer have the same etiology as leukemia, colon cancer, or prostate cancer? People exposed to pesticides, toxic chemicals, or chronic stress can develop cancer.

The Bubble Boy, David

Regarding thalassemia, researchers do not know if it is caused by a mutation, the deletion of genes, or a histone modification. They do know that people with thalassemia do not produce enough globins to carry the vital element oxygen to all parts of their bodies. Other than the thalassemia, my family has been healthy. My father and my mother lived to be almost a hundred years old. There is a movie that pops up in my brain at times—*The Boy in the Plastic Bubble*. It was inspired partially by the life of a boy named David who was born without an immune system. His body was unable to make antibodies to destroy intruders. David lived in a sterilized environment (the bubble) until he was twelve years old. A bone marrow transplant from his sister did not help. A virus was present in the cells she donated to him, and it killed David. He did not have T and B cells to alert and organize a host of auxiliary killer cells that would protect him from viruses.

The literature I researched claims no linkage between thalassemia and leukemia. For thalassemia, gene therapy may offer a solution. Trials with humans have not produced the desired response, but we were not born flying. Whether we include induced pluripotent stem cells (iPSCs), embryonic cells, gene therapy, microRNA, or cloning, we are definitely going in the right direction. We will convert blood stem cells into T cell progenitors that will join B cells in the battle against pathogens.

My initial concern over the etiology of leukemia was partially answered during an international symposium in Germany in 2004: "Several studies reported the derivation of leukemia stem cells from hematopoietic stem cells as a result of epigenetic events ..." [24] Whether in the United States, Europe, or Eastern countries like Japan, South Korea, or China, embryonic stem cell research, as well

---

[24] *New York Academy of Science*, 1X

as induced pluripotent stem cells, will be in the forefront of medical research and pharmacotherapy. With iPSCs, we can create any cell in our body. It is even possible to build tissue and organs for a body in need of repair. C. Venter's synthetic bacterium may become a handy repair tool.

The most beautiful part of iPSC technology is that the problem of rejection by our immune system will be overcome, because we will be using stem cells from the same person in need of repair. Our organism is a factory of functional cells. Japanese and American scientists are at present the leading researchers in this field. Making use of these technologies, we will not be doing anything that is out of this world. We have a natural tool factory inside us. Watson and Crick placed us on the road to decipher the human genome code. Now we can enhance our brain to absorb and intelligently use all this technology to improve and accelerate brain cells, which have moved us ahead of other hominids.

Learning from Sequenced Genomes

WHAT CAN WE LEARN from Watson's genome sequence and the millions of small-nucleotide polymorphisms that are attached to the three billion pairs of the genome? James D. Watson's genome sequence showed about 3,300,000 SNPs, or simple substitutes of one base for another at particular sites in the genome. Watson's genome variation, as well as that of Craig Venter, seems to be typical of a genome variation of individuals of predominantly European ancestry. Of the 3.3 million SNPs, 10,654 caused amino acid substitution within the coding sequence. In addition, the leading author of this article, David A. Wheeler, wrote that his team identified copy number variations resulting in the large-scale gain and loss of chromosomal segments ranging from twenty-six thousand to 1.5 million base pairs. For an untrained individual alien to genome sequencing, these are a lot of substitutions, gains, and losses of base pairs. For those unaccustomed to seeing so many changes within a single genome, it might be a little scary. However, the high number of variants did not affect Nobel Prize winner and brilliant scientist J. D. Watson mentally or emotionally; he is mentally a well-rounded individual.

We have the genome sequences of two more brilliant scientists: C. Venter and Francis Collins. I have not read enough about the genomes of these scientists, but I would not be surprised to see similar patterns in them. M. V. Olson, commenting on Watson's genome, wrote, "The genome sequence seems to show that he is a carrier for a handful of mutations … but these mutations have no known effects on him, wrote V.Olson." Watson, along with F. Crick, gave us the double-helix structure over half a century ago. Now these brilliant and remarkable men have left us their genome sequences so that we may learn about ourselves through millions of gene alterations found in their genomes. I will close Olson's lesson with another quotation from him: "The challenge in human genetics now is to learn how to correlate genotype with phenotype (the external appearance) with special attention to disease predisposition and response to therapy."[25]

DNA, SNPs, and Diseases

A brief comment on the above is not out of order here. Companies that advertise their ability to decode your genetic code and reveal to you where you got your beautiful legs from or how to change the color of your eyes are publicity seekers aimed at making profits rather than achieving scientific accomplishments. However, there are sophisticated and reliable companies that perform genome scanning and reveal to you potential assets and liabilities in your genome so you may plan accordingly.

Again I offer a word of caution: interpreting for you the potential assets and liabilities that may number in the millions is unsurprisingly not always accurate. Among those genomic markers are SNPs (short-nucleotide polymorphisms). They are DNA variants that in some cases are associated with your susceptibility to a disease. But of course, the mere presence of these SNPs in your DNA, perhaps in the wrong place, does not mean that you will get the disease. I

---

[25] *Nature, Vol,* 452 (April 17, 2008): 819–20, 872–6.

would like to remind you once more that both J. D. Watson and C. Venter's sequenced genomes show liabilities; however, generally, both are healthy and brilliant scientists. Both have been blessings to humanity.

Take, for instance, variants of a gene dubbed *APOE*, which is associated with Alzheimer's risk. The presence of SNPs can form APOE variants that contribute to the degeneration of neurons, ultimately leading to Alzheimer's. How are introns, the noncoding genes, and their variants related to the disease? How do exercise, meditation, and the maintenance of a mentally busy and well-focused brain, like those of Venter and Watson, influence the growth of these variants into liabilities or assets? Why are these companies that interpret the propensity for a specific disease when there are so many epigenetic and environmental causal factors jumping on us 24-7? Besides, a good number of these findings are based on statistical averages, and there are a significant number of samples that do not fit or fall within the average category. I do not know for sure, but if we were to go by an average sample regarding scientific talent, for example, John Nash, C. Venter, J. D. Watson, A. Einstein, and even da Vinci would have to be counted out. I am not throwing out the average scale. What I am saying is that before you lose sleep and money over your DNA variants, please pay a visit to your physician or a qualified health professional.

Genome Surprises

Genome sequencing has brought us many surprises. We have sequenced a few animals, including humans of European, African, and Chinese ancestry; this has provided us with insight into our own evolution and development. Among our surprises was the incredibly large amount of noncoding genes taking up space inside our skulls. We have learned that a mutation—a change of a single letter of DNA code—is not that bad at all. During gene duplication, some genes will be ill prepared to function normally and will be degenerated. The cell has its own self-correcting mechanism. However, a produced mutation may occasionally slip by and allow a duplicated gene to prove itself by doing new and useful jobs for the good of the cell and, ultimately, your organism. A gene may strive to serve the cell and may give rise to competition of functions. Smell and sight in dogs and birds might well illustrate this point. Do not forget that shape determines function. Many things take place during cell specification that I cannot elaborate on in this essay. Here I am not referring to the battle of the fittest, but rather to which shape fits in and best serves the cell and organism. "We humans have about 400 genes for smell receptors alone, all of which derive from just two in a fish that lived around 450 million years ago."[26] This is not a short trip by any means, but there is not a way we can compare our senses of sight and smell with those of an eagle, a hawk, or a dog.

During the last decade, we have been looking deep and close into the intricacy of the genetic code of life. Letters change within a gene, misplaced codons and introns enter promoters' domains, and DNA and histone modifications can interfere during protein synthesis and embryo and fetus development. Consequently, all these steps may provoke malformations and diseases. Even a complete chromosome can be duplicated, as happens in Down syndrome.

In 2008, biologist Wen Wang in China, working with several species of fruit fly, an excellent research tool, was able to identify new genes that have evolved over millions of years. To his surprise, he found that 10 percent of the new genes had arisen through a process called retroposition. Retroposition occurs when messenger RNA copies genes; this is the RNA template sent to the protein assembly factory—the ribosome. These retroposition copies of genes are turned back into DNA and are inserted somewhere else in the

---

[26] *New Scientist* (November 22–28, 2008): 44.

genome. The interesting thing for us in this discovery is that the gene copies created by retroposition are not the same as the original ones. We know that genes consist of more than just the sequence coding for a protein. Among laborers involved in gene and protein synthesis are promoters that send signals for additional participating molecules.[27]

Promoters, as gene regulators, are something we have to reconcile. They are involved in determining the location, time, and volume of the protein to be made. Promoters are facilitators that we need to fully employ on our side and convert into first-class research and therapeutic tools. What is needed is a tool to eliminate most junk DNA. In addition, it would be best to place the limbic system under the strict control of your prefrontal cortex. It may also be prudent to put together a synthetic pathway between the hippocampi and long-term memory storage loci so memory storage can be turned on and off at will.

A synthetic gene in a chromosome, as far as we know, is science fiction at the present time. However, gene duplication seems to be as common as mutation. During gene duplication and reshuffling of its component parts, the resulting new gene may well be different from one's own ancestral pool of genes. Genes are the instructions to make proteins, and proteins are the basic pieces of matter nature made necessary to build your body. Furthermore, genes hold the secrets of your entire life ever since you existed as a simple organism. By studying your genes, you will learn about all the secrets your own ancestors did not share with you.

The Last Ten Years in Genome Sequencing

A layperson overview of genome sequencing over the decade ending in November 2011 would include speculation about the number

---

[27] *New Scientist* (November 22–28, 2008): 46.

of protein-coding genes in the human genome, ranging from over one hundred fifty thousand down to about twenty-two thousand. Junk DNA, epigenetics, and probably human DNA's similarity with that of other mammals, such as the mouse and chimpanzee, would not be considered. However, Professor E. T. Dermitzakis, from the Department of Genetic Medicine at Geneva, Switzerland, points out that in the genome, the exact location of regulatory regions was unknown during this time. Only a small fraction of variations that exist in the human population had been characterized.

The basic components in each genome are largely the same, but the way they are used differs from tissue to tissue and from person to person. Understanding the rules of gene regulation—the grammar of the genome—is key to the understanding of the human body.[28] Professor Pardis Sabeti from Harvard University said, "During the pregenomic era, revolutionary genetic was a painstaking process. Scientists hypothesized instances of selection and sought confirmation case by case." As of the year 2000, only a handful of cases had been identified. Technological and analytical advances in the past decade have enabled us to progress from hypothesis-testing to hypothesis-generating science. We can now scan the entire genome to identify variants that come about by way of natural selection. [29]

Chief Scientific Officer Eric Steps from Pacific Biosciences of California said, "The first human genome sequence, published in 2001, provided a canonical reference from which to understand genome structure, as well as registry of functional units … At present, thousands of genes that influence susceptibility to hundreds of diseases- associated phenotypes has been identified."[30] Science fiction is fiction until we learn to do it.

---

[28] Science, Vol. 331 (February 11, 2011): 689.
[29] Ibid., 690.
[30] Ibid., 691.

The Watson, Crick, and Venter Genomes

IN GENERAL PARLANCE, J. D. Watson's genome sequence showed us that his biochemistry served him well. Watson left the proteins, amino acids, codons, genes, DNA, and RNA that we have been talking about in his genome for us. They came from a person dedicated to working for all of us. His work produced a brilliant scientist who united his chemistry with another jovial and equally smart person, F. Crick. Crick's chemistry connected very well with that of Watson. Neither of the original players in the discovery of the double-helix structure stuck together as Watson and Crick did. I wonder to what extent Watson's SNPs and mutations—all parts of the dynamics of our chemistry—played a role in his behavior and stamina to stay on course. To succeed, he ignored roadblocks he encountered during his challenging life.

I remember the biography of another outstanding scientist of the nineteenth century, Madame Marie Curie. She defeated prejudice and male chauvinism and remained loyal to her ideals and love for science. Her chemistry will endure for centuries to come. She died doing research in her effort for us to understand ourselves and live happy lives. Her laboratory was her universe, and everything in that universe was subject to scrutiny and analysis. Cause and effect were the golden rule for her research.

Today we do research with tools that can be observed and replicated in order to make use of our laboratory findings for the benefit of all human beings. One of our most widely used research models in medical research is the mouse. Most recently, in 2011, an orangutan's genome was sequenced and shown to be 97 percent identical to that of humans. We already know that our next closest relative among hominids, the chimpanzee, is 98.4 percent like us, humans. You may wonder how the sequencing of the genomes of these animals helps us medically and otherwise. Just think about chromatin methylation and modification, as well as other relevant epigenetic causal factors in human illnesses. They may be identified through the sequencing of the genomes of animals with genomes similar to ours. For instance, when we silence a critical gene in a chimpanzee, we learn how it affects human health and behavior. Consequentially, we can develop therapeutic medical tools to either correct the gene and its proteins or arrest the disease.

Gene Targeting Is Not Sci-Fi

Gene targeting does not belong in science fiction television shows intended to get your adrenaline running high in your body. Our preferred animal of research model, the mouse, has the same organs as humans do, and their genes are about 95 percent identical to those of any male or female specimen of the human race. Using gene-targeting technology, scientists the world over can knock out—meaning "turn off," "suppress," or "totally inactivate"—a gene in a mouse, mimicking a human disease. This medical technology can be applied in treating real human diseases. Scientists can force a mutated gene to reappear in specific organs and tissues in our bodies, including our brains, and observe its behavior as it relates to a particular illness. The knockout technique has provided genome researchers—particularly medical researchers—the opportunity to identify and observe each gene in the human body from birth to death using the mouse model. Our most recent estimate reported that we have about twenty-four thousand genes in our genome.

You may argue that with our modern technology, learning everything about our genes should not be an insurmountable challenge for us. You may add that identifying a disease risk marker in our genome seems to be easy job. This is exactly what we thought when we first started searching for dysfunctional genes. The problems we did not anticipate were the role of junk DNA, DNA methylation, histone modifications, and, in general, all epigenetic and environmental agents contributing to the etiology of any disease in the human body. There are millions of tiny units in the human genome whose functions are unknown. These units and functions are like uninvited guests with weird behaviors. Making this issue even worse, they have made our bodily organs and tissues their permanent home. But thanks to our brains, we are determined to get to the mountaintop and take control of our destiny and health.

The Science King of the Twenty-First Century

Nanotechnology, our most recent brain child, has engineered tools not only to look into and possibly repair dysfunctional genes but also to go into molecular and atomic levels of research. It is the science king of the twenty-first century. Unless we use it against each other, we will change the way we see ourselves and the world around us. In a quarter of a century of research, we went from a computer that needed a truck to move it around, to something as small as your wristwatch. We have discovered materials stronger than steel but lighter than a feather. Solar energy will be used to energize traveling miniature vehicles carrying human cells containing memory-carrying proteins. Cells will replicate at sites chosen by humans on planet Earth. Hibernation will be used as an adjunct technology during voyages.

This technology is not in science fiction books written for recreational purposes; it is in our laboratories. The anticipated problems do not lie in our brains; they lie within our economic, social, and political systems around the world. Our greatest challenge is learning how to feed and control around 10 billion people on Earth within the next forty to fifty years. A recent article in a prestigious scientific journal wrote that feeding 15 billion people by the middle of the century would be "easily possible" if more land were used for food cultivation. However, that land is located in Africa and in Latin America. Not mentioned in the article is that land in Latin America and in Africa that in the past was used to feed people is now being used to produce fuels for the luxurious cars, yachts, and airplanes of the developed world's upper class.

Communication among people will be much easier and more accessible to the majority of the population in any given country. Migration around the world will be an enormous problem that will be most difficult to control. In addition, the world areas slated for future agricultural use are exactly the places where very rapid population growth will likely occur. India and China and countries between those two possess about a third of the world's population. Many inhabitants of these nations will be moving out of their national boundaries rather soon. There will not be a power strong enough to challenge it. This is already occurring in many parts of the world. Western Europe seems to be inundated with migrants from Eastern and African countries. They arrive in homemade boats, on foot, in trucks, and by posing as tourists.

There are justifiable arguments that we will come to some sort of agreement for control of unrestrained birth. It is a little over a half century since we established the United Nations to avoid wars. However, many wars have been fought since the UN was established. The big boys at the UN are pursuing their hidden agenda. Now, imagine how we can get together to solve the problem of putting food on the table for billions for people? Agreement on a single subject has not been a common occurrence in the history human behavior. Man spends more money on defense weapons (which actually is nothing more than a weapon to kill his neighbor) than food, health, and housing.

The US Department of Energy has launched a late technology initiative to convert sunlight into hydrogen and other fuels. Its director, Nate Lewis said, "We have to scale up from nanoscale to macroscale." I hope it will capture photons and produce electricity, which can be used to split water molecules and produce hydrogen. With artificial photosynthesis, photons from the sun could drive a wireless chemical conversion process to generate fuel.[31]

Converting land that has been dedicated for millennia to producing rice, corn, wheat, bananas, etc. into areas of fuel production for the super-rich is not the solution to the above-mentioned problem. The answer already exists, and we can thank our brains for it. Genetic engineering of seeds, animals, fish, and plants has been among us for millennia, but at present we can do it faster and more productively for all of us. We can engineer seeds to fight insects and survive during drought. Instead of one crop a year, we can engineer seed that will produce two crops a year. We can also increase the sizes of fruits, vegetables, and roots. Most such innovations have come about after the double helix was discovered and researchers began sequencing genomes from animals and plants. Genetically engineered corn and soybeans have been two of the greatest accomplishments in making use of DNA technology. Conversion of fertile land to provide fuel to move vehicles for the rich is not going to solve our world problems. The sun is the greatest source of energy we need to explore. And we can also extract hydrogen from our oceans. We will have to invest in research with little initial profit, but the final solution will save us all.

---

[31] *Nature*, Vol. 466, no. 7306 (July 29, 2010): 541.

Research in RNA, Amino Acids, and Proteins

RESEARCH IN DNA, RNA, amino acid chains, and proteins has kept minds focused for many years. The double helix can even be found in food boxes. The way proteins are made was a curiosity we could not put away or ignore even during sleep. Researchers helped us during this time by discovering the basic RNA molecules involved in protein synthesis. I previously mentioned that rRNA, tRNA, and mRNA are all involved in the making of a protein. Small interference RNA (siRNA) is a molecule that becomes a nuisance when it interferes during the synthesis of a protein molecule. We all know that proteins are peptide chains.

The above-mentioned ribonucleic acids and molecules play a significant role carrying and transporting component polymers from the cell nucleus to the final assembly station—the ribosome. Recently, Phillip Kapranov and others, in the journal *Nature* (July 29, 2010), suggested a novel RNA-copying mechanism. He claims that RNAs are not merely degradation products of longer transcripts but could indeed have a function. The profiling of RNAs, including the sRNAs, not only can reveal novel transcriptions but also can make clear predictions about the existence and properties of novel biochemical pathways operating in the cell. This article appealed to me because, among other things, I have maintained that RNA is the human body's mechanic that puts the parts together after receiving a message or order from the cell nucleus. RNA folds itself into many shapes and carries out multiple functions, which has made us show more respect and admiration for this acid than ever before.

Some readers may think that I am overly concerned about siRNA as a "spoiler" in protein synthesis. Their concern might well be justified. However, I invite you look at pharmaceutical and investor corporations that want to convert siRNA into a therapeutic tool to take care of some of our illnesses. RNA interference was discovered in 1998, and in 2006, the group of scientists who discovered it were awarded the Nobel Prize. Soon thereafter, Merck, the giant pharmaceutical company, paid more than $1 billion in its competition to make drugs out of siRNA. Another giant in the drug business, Roche, gave up in this competition after spending more than $500 million on the technology. Furthermore, please read Heidi Ledford's words for *Nature*: "The development of RNA i based drugs has stalled as companies confront the challenge of delivering RNA molecules, which are notoriously fragile to target cells in the human body, and then coaxing those cells to take up the RNA."

Michael French, chief executive of a biotech company in Bothell, Washington, added, "Getting these molecules exactly where we want to go is a little more difficult than originally thought."

Equally curious and interesting for me is the body's building block, the protein. There are some proteins and their receptors that get a lion's share of researchers' attention. Naively, our attention has been geared toward our brain cells and membrane receptors. Heidi Ledfor writes, "About one quarter of approved drugs target members of a single protein family, the G protein coupled receptors ... Members of this clan- the largest protein family in the human genome, control everything from hormone signaling to the perception of light and scent. A no small amount of money from the USA., $290 million will be re-directed from learning amino acids folds that comprise a protein ... to solving some of the world most troublesome and medically relevant proteins including the G protein coupled receptors." Besides this protein, mitochondria as well as proteins that regulate gene expression will be principal targets. I feel relieved by this announcement. The focus will be on proteins—specifically, proteins that are medically relevant. I want to know not only how proteins are formed but also how they relate to and become the causal factors of many of our health problems. You already know that protein packaging provides shapes, functions, as well as the length of the chain of peptides. In addition to multiple containing subunits, it takes on various roles in cells, organs, and tissues of our bodies.

A quick word on protein synthesis is necessary before I move on to the next paragraph. The dogma among biologists is that a protein needs a structure to be able to function properly. Every book I read emphasizes that function demands structure. There have been serious arguments that even the most unstructured environment is ruled by some type of structure and order. Our personal opinion is in consonant with structure, function, and order. Peter Wright, a protein biophysicist at the Scripps Research Institute in La Jolla, California, told a group of scientists attending a meeting of the American Association for the Advancement of Science in Washington, DC, "The recognition of disorder has grown dramatically." Data are fast accumulating from all fronts—biophysics, bioinformatics, and cell biology—in support of widespread disorder. P. Wright and team member J. Dyson wrote a review back in 1999 pointing to the growing collection of proteins that seemed to function despite their disordered states.

However, there are some researchers with broad imaginations, such as Joel Janin from Gif-sur-Yvette in France, who says that "the whole concept of disorder seems incompatible with the lock and key model. You might as well try to open the door with cooked spaghetti. Little by little, a fundamentally new structure of relationships between protein sequences, structure and function is emerging. "It consists of a continuum running from the most rigid lock and key enzymes and molecular machines at one extreme through to durably unstructured spaghetti, spanning all degrees of structural ambiguity in between."[32] How this cooked spaghetti structure will work or is already functioning, is an enigma to me, but I will keep an open mind and see how it turns out. When Rosalind Franklin, J. D. Watson, and F. Crick were attempting to put the double helix together, doubts seemed to outweigh good intentions.

G Protein Receptors

What are G protein-coupled receptors? These are a large group of proteins that are attached in the plasma membrane of most cells. They seem to look for and attach to agonist molecules responsible for conveying sensory information. They are also involved in the transmission of changes in physiological states of the cell. Once one of these receptors is activated, a G protein initiates a signal that controls multiple biological processes. Determining how best to approach this challenge is going to require elaborate teamwork. The mouse and the petri dish are excellent research tools, but translating findings to the physiology of a protein receptor is another issue. The cell's internal aqueous environment is further complicated by internal chemical reactions. Likewise, the cell's outside is equally influenced by an aqueous environment. Protein shape and function are conditioned, among other things, by the aqueous environment. The way water molecules come in contact with proteins in various cells affects protein functions and related medical problems. The dynamic exchange of water and protein atoms and molecules in

---

[32] *Nature,* Vol. 471 (March 10, 2011) 151–3.

humans at different stages of development is another challenging issue. This huge family of protein receptors is responsible for many functions I am not equipped to fully elaborate on. However, the chance to research the most recent and relevant literature available is something I could not avoid. This is not a local issue but a worldwide challenge.

A G protein-coupled receptor is essential for Schwann cells to initiate myelination. Signals from the axon activate expression of specific transcription factors, such as Oct-6 and Kro-20. I write these names and numbers just for reference purpose, not to bug your brain. These two molecules begin myelination in Schwan cells. These are glial cells that produce myelin in the peripheral nervous system. Myelin is a lipid, or fat, that is wrapped around the axon. It insulates the axon and allows the electrical message to reach its target fast and uninterrupted. Lack of myelin in the neuronal axon can be fatal. It may provoke a serious illness, such as multiple sclerosis, amyotrophic lateral sclerosis, and Krabbe disease. A type of brain cell called a glial oligondendrocyte produces myelin. In your brain are cells known as neurons. Neurons are composed of a body, which looks like a round disk with many fibers extending out. These extensions are called dendrites, and they receive messages from other brain cells—in particular, neurons. There is another type of or fiber coming out of the cell body called an axon. Messages from the neuron nucleus go through the hillock and into the axon. A hillock looks like a ring linking the cell body and the axon. The myelin covers the axon, similar to the way plastic is used to cover electric cords. But there is a difference between the axon covering and the plastic covering on an electrical cord. First, the axon is not totally covered by myelin. There are small segments on the axon that are free of myelin. These spaces allow electricity in the axon to gain speed by jumping from one node to another down the end of the axon. This electrical impulse will provoke a chemical reaction, meaning a neurotransmitter will be released at a synapse, which is a space between neurons in the brain.

Among the neurotransmitters released into the synapse are serotonin and dopamine, which I mentioned in the preceding pages. Loss or destruction of myelin in the central nervous system is most common in the brain, spinal cord, and optic nerves. Multiple sclerosis is the primary disease of demyelination. Some of the known diseases caused by destruction of myelin are multiple sclerosis, Krabbe disease, Tay-Sachs disease, phenylketonuria, Hurler syndrome and neuromyelitis optica. Of the above diseases related to myelin loss, I have had professional experience with multiple sclerosis and Krabbe disease. Multiple sclerosis is characterized by recurrent episodes of demyelination. It is a chronic autoimmune inflammatory and neurodegenerative disorder of the central nervous system. The durations and frequency of the attacks do not seem to be well established as yet.

G protein-coupled receptors are responsible for the majority of cellular responses to hormones and neurotransmitters as well as the senses of sight, olfaction, and taste. The ß2 adrenergic receptor has been a model system for the large and diverse family of G protein-coupled receptors. During the last ten years, sequencing of the human genome led to the identification of over eight hundred GPCR genes. The B2AR-Gs crystal structure complex provides the first high-resolution view of transmembrane signaling for a GPCR. Rasmussen and others have a framework to design experiments to investigate the mechanism of complex formation, GTP binding, and complex dissociation.[33] Dies Meijer, writing for the journal *Science*, offered the following: "Myelin is laid down and maintained by dedicated neuroglia cells- oligondendrocytes- in the central nervous system and Schwan cells in the peripheral nervous system."[34]

---

[33] *Nature* (September 29, 2010) 549–55.
[34] *Science, Vol,* 325, no. 5946 (September 11, 2009), 1353.

My experience with Krabbe disease was a very sad one. An eighteen-month-old girl, otherwise considered healthy by the local pediatrician, began crying without any obvious provocation. Her mother took her to the local physician, but to no avail. She was later taken to a specialist who diagnosed her with Krabbe disease. She was later transferred to a well-known pediatric hospital on the US East Coast. Her initial symptoms were that her feet turned inward, she was unable to stand, she could not walk, and she complained of pain all day long. Soon afterward, the little girl began to lose her sight and hearing. After a few weeks in the hospital, she was discharged home to her mother's care. She is now in a wheelchair and receives weekly nursing visits and undergoes monthly physician checkups.

Another reliable resource on myelin and Schwan cells wrote, "The axon of many vertebrate neurons are enclosed by a chain of supporting cells called Schwan cells in the peripheral nervous system, that form an insulating layer called the myelin sheath."[35]

A Personal Encounter with Histamine

During my clinical practice, I have seen many people with rare symptoms and diseases that have strengthened my dedication to assist individuals in pain. One area that always challenged my ability to help anyone is allergies. We have been fighting allergies for a long time. Dust, pollen, weeds, cats, dogs, foods, and even an occasional rare perfume can provoke wheezing, a runny nose, and watery eyes in an allergy sufferer. A visit to a physician usually brings relief after the prescribed medication is taken. However, the prescribed medication does not always work. One should not blame the medication or the physician for this. Although humans are made of twenty amino acids that will be formed into proteins, the end product of amino

---

[35] Neil A. Campbell, *Biology*, 3rd ed. (New York: The Benjamin/Cummings Co., 1993), 984.

acids and proteins—our bodies—are not entirely equal. We end up being two unique individuals who respond to internal and external stimuli in different fashions.

However, despite our differences in behavior, there is always an average response. And a medication, or rather the usefulness of any medication, is measured based on the average individual, not on the difference between each individual person. Consequently, a medication that worked wonders in Peter will not necessarily perform the same miracle for Joe or John. Remember: we carry our genes from our ancestors, but we are also greatly influenced by epigenetics and our immediate environments. My mother and two of my siblings were always complaining about allergies. Allergies have been my inseparable companions for many years, along with my relatively poor digestive system. Greasy foods like pork chops, milk, cheese, and even butter pecan ice cream (which makes my mouth a lake of saliva), do not sit well inside my stomach. Somehow those enzymes responsible for breaking down fat are not helping me. I avoid those foods, although my eyes and an internal voice call me a coward for ignoring my desire to enjoy myself. I buy appropriate enzymes from a health food store.

Besides allergies, I have learned about another health nuisance—histamine. I have said that the human body is a very complex chemical reaction factory producing millions of reactions that we are committed to fully understand. You are correct when you say that many of those chemical reactions are the effectors of some of our diseases. Health foods stores and pharmacies have full shelves waiting for you to choose from many similar remedies. To calm down my stomach acid and runny nose, I carry with me the antihistamine called Diphenhydramine.

Histamine as a Chemical Messenger

Histamine is a chemical messenger that is responsible for handling a wide range of cellular responses. I have been talking about DNA and RNA molecules going through multiple reactions until proteins come out to form organs and tissues. Take a minute and visualize all of the chemical reactions going on at this precise moment inside your body—specifically, inside cells. Among the reactions I am referring to are allergic and inflammatory reactions. It is likely you have had an allergic reaction of some sort at some time. What is not visible and difficult to explain is the inflammatory reaction taking place inside you. Some asthma attacks are caused by an allergic reaction in the lungs. When suffering an asthma attack, along with the inflammation in your lungs, you may have an excess of mucus that makes it very difficult for you to breathe. You might feel as though you are choking on your own saliva and mucus. You might cough again and again to clear your throat and breathe freely. In addition, you might find that there is an excess of gastric fluid in your stomach. As you try to cough out all the mucus coming out of your lungs, your stomach might produce a full cup of vomit.

All this pain and discomfort during an asthma attack is basically provoked by histamine or an excess of it. This chemical messenger mediates a lot of responses in millions of cells in your body. It is suspected that it works in brain cells too. Besides all the problems that an excess of histamine can create for you, it is a powerful vasodilator.

Serotonin is highly involved in brain functions. It is a neurotransmitter involved in depression. However, the largest amount of serotonin is found in cells in the intestinal mucosa. "Serotonin has multiple physiologic roles, including pain perception, normal and abnormal behavior, including affective disorders, and regulation of sleep, temperature, and blood pressure."[36] In addition to the well-known serotonin, we have another public friend that most people talk about, sometimes even over the dinner table—dopamine. This neurotransmitter is a chemical that makes you feel extremely happy when produced in excess quantities. It is also involved in two chronic and degenerative brain diseases: Parkinson's disease and schizophrenia. In Parkinson's disease, cells producing dopamine in the substantia nigra die out. Recent medical technology is helping people with Parkinson's disease. There are some genes involved in schizophrenia, but this disease is more of a syndrome than a single encompassing behavior. There are effective medications available at present to control most of the symptoms of schizophrenia. Symptoms of schizophrenia are not difficult to identify by health professionals. Youngsters from ages sixteen to twenty-one are very vulnerable to strong psychological and physiological stressors. Individuals with a schizophrenic predisposition need extra help from parents and teachers as soon as possible.

---

[36] P. C. Champe and R. A. Harvey, *Biochemistry*, 2nd ed. (Philadelphia, PA: J. B. Lippincott, 1994), 265.

E. Kandel's Research

DR. ERIC KANDEL AND Eleanor Simpson, both neuroscientists at Columbia University in New York City, reported in 2006 a discovery made on D2 receptors associated with schizophrenia. Dopamine D2 receptors were implicated in the etiology of the disease. These two scientists used engineered mice to mimic deficits of short-term memory and attention in schizophrenia. The schizophrenic mind loves to travel; it jumps from one theme or subject to another, unable to focus in the present. The schizophrenic brain's capacity for hallucinations and delusions seems to be a bottomless pit even during catatonic states. The dopamine D2 receptors suspected of contributing to schizophrenic symptoms are located in two regions deep in the brain. These two clusters of neurons are commonly known as striata and consist of putamen and the caudate nucleus. They are part of the basal ganglia, which are associated with another chronic and degenerative disease called Parkinson's disease. Dr. Kandel and his team engineered mice with at least 15 percent more D2 receptors in the striatum than normal.

The researchers found some deficits to working memory. Working memory is memory of short duration; it is there for you to handle a task at hand. Synaptic sensitization and permanent anatomical neuronal changes by protein synthesis do not fall into this group of memory. The mice in question took much longer than control mice to master the task or game during the experiment. Control mice were not engineered; their D2 receptors were not touched. The researchers concluded that extra D2 receptors cause permanent brain damage. Dr. Kandel added that such damage probably occurs while the brain is still developing. This experiment dealt with schizophrenia-related symptoms relating to learning and memory only. It did not deal with hallucinations, delusions, mood changes, or erratic and violent behavior. I will follow up on this experiment later in the book.

An interesting observation I would like to point out is the connection to the developing brain. Other researchers have followed cases of attention deficit hyperactivity disorder (ADHD) that have turned into full-blown schizophrenia. In addition, parents, teachers, and mental health professionals dealing with children can observe early symptoms. Observation and identification of early symptoms can contribute to early diagnosis and treatment for schizophrenic patients. During early adolescence, while the prefrontal cortex is going through immense changes, dendrite growth and maturation may relate to Dr. Kandel's observation that the changes he noted occur while the brain is still developing.

Dr. Solomon Snyder at Johns Hopkins Medical Institution in Baltimore concluded that the mice findings reported in the journal *Neuron* could help researchers develop new drugs to prevent such early brain damage from occurring in people who may be susceptible to schizophrenia. Whether I like it or not, luminaries like John Nash, Salvador Dali, and van Gogh popped into in my brain. They contributed so much to humanity despite their inner pain; we would be a lot poorer without them. Schizophrenia has been an enigmatic disease for millennia. It has been blamed as a punishment from the heavens, and a curse from relatives and friends. Schizophrenics have been ostracized, imprisoned, burned at the stake, and even guillotined.

Most of our medications for schizophrenics are aimed at controlling brain chemicals, such as dopamine and serotonin. The big problem we face is that it is a syndrome rather than a single isolated symptom that can easily be identified. The schizophrenic patient can go from having a completely flat affect to being emotionally hyperactive. While the patient has a flat affect, he or she may be verbalizing the most horrible accident his family has experienced without conveying an ounce of emotion. During therapeutic sessions, the therapist may wonder if the emotional part or parts of the patient's brain were surgically removed. However, the same person seemingly without emotions may become agitated or emotionally aroused by a sensory stimulus or a past episodic memory and present a seemingly normal affect. His mood may unpredictably change from time to time. Historic records show that we have had the disease with us ever since the beginning of written language.

Alois Alzheimer and Spaghetti Strings

Before I proceed in discussing proteins, I would like to write down a few words about the most dreadful dementia of all, Alzheimer's disease. Alzheimer's disease was first fully described by Alois Alzheimer in the nineteenth century, and it continues to be as frightening and elusive as ever. We know that proteins shaped like spaghetti, peptide chains known as plaques, and the tau proteins are, as far as is known at the present time, the two main culprits behind this degenerative and chronic brain disease. This disease robs you of your personality because it destroys your memories. The plaques mentioned above block communication between neurons, making synapses disappear.

Another destructive scenario takes place when the neuron's internal organelles or fibers break down and form an entanglement that collapses the neuron. The complex and harmonious functions of each neuron cease to exist. What is left of the victim is a body with a head, but the brain inside the head is no longer generating chemicals and electrical impulses. The neurons that made it possible for you to do complicated and elaborate tango steps, play the piano, do surgery, or appreciate Picasso and van Gogh paintings are not available to you anymore. Even when sweet voices from loved ones whispering into your ears, there are no receptors in your brain cells to prompt you to respond to and reciprocate the sweet feeling. This disease is generally associated with old people, but it has been seen in patients in their forties. It is not a disease of only old people after all.

In 2006, William Jagust, a physician and neuroscientist, said at the time, "Many people believe that we are going to have treatment for Alzheimer in five years." Many miles away, in Sweden, a group of researchers tested spinal fluid for concentrations of tau protein and beta-amyloid, both of which are strongly implicated in Alzheimer's disease. The group tested consisted of 137 people from fifty to eighty-six years old. Patients whose spinal fluid at the beginning of the test had abnormally high tau protein and low beta-amyloid concentrations were nearly eighteen times more likely to develop Alzheimer's disease than those with normal concentrations. Those researchers had hope that there would be effective drug treatment by 2010. There is a lot of confidence among scientists, especially in light of the most recently developed technologies. An effective treatment, in my opinion, is not far away. We are joining hands and resources to defeat this disease.

Following on Proteins

No other molecules in the living world, except DNA, are as chemically, structurally, and functionally complex as proteins. Although the cell nucleus sends out the message for protein synthesis, the message itself (meaning the selection of specific nucleotides that will recognize a complementary molecule with an amino acid to be pasted together at a ribosome) is conveyed while other chemical reactions are occurring in the cell's cytoplasm, organelles, and tissues; this is an awesome engineering accomplishment. Crucial among choices to be made during this chemical wonder (if they are choices) are selections of codons, appropriate amino acids (often within a group of six triplets), amino acid groups, carboxyl groups, side chains, shapes, and functions, and sizes. How many peptides for a specific protein function? How do proteins respond to multiple internal and external stimuli and still maintain their integrity and function?

The human body makes many protein molecules with highly specialized functions that are vital for the survival of every living cell on our planet. Imagine a protein that exists in a cell or organism in a dry desert and compare it with a protein in a cell in the coldest region in Siberia. Consider a protein that is involved in generating electricity in a brain cell and one that is in the toe of your right foot. There are proteins that are turned into highly skilled soldiers of our bodies, such as the T and B cells of the human immune system. I have proteins that are at least over eighty-six years old and still kicking. Most of my brain neurons are functioning pretty well. Imagine how many hundreds, if not thousands, of different types of proteins are doing specialized jobs in your body at the present time while you read this book. Each one of these proteins has its own particular amino acid sequence. We must also remember that each amino acid has particular properties. There are polar and nonpolar amino acids that are decisive in protein folding.

From Mind Speculation to Brain Research

Just a couple of decades ago, mental health practitioners viewed brain disorders as just mental problems. The issue I want to emphasize here is that "mental" is considered as having no matter. The mental realm and psyche were, and still are, considered in some quarters as something separate from anything having flesh. Because these practitioners considered mental problems something that could not be linked with to a brain tissue or organ, no loci in the brain were considered specifically responsible for any brain disorders in particular. Psychologists ruled out brain tissues and organs as the subject of research and exploration in their efforts to find a reliable and trustworthy cure for brain disorders. They reasoned that if mental problems did not have biological etiology, they had to be caused by environmental factors. They were not too far off in their efforts to find a culprit for psychiatric diseases and disorders. However, they chose patients' homes as the most relevant factor leading to brain problems. And at home, all the blame for all types of behavioral problems was attributed to the mother. Second in line in the culprit list was the father, followed by family disputes, poverty, etc. I recall some mental health practitioners blaming the mother for being "an emotional refrigerator or an emotional armadillo" and thus for being responsible for her child's autistic, schizophrenic, or schizoid behavior, ignoring a biological base for those brain diseases.

Nowadays, scientific approaches based on genomic studies and neuroscience, as well as all biology-related medical diagnostic tools—such as PET, FMRI, and similar technological medical tools—are changing how we look at and treat psychiatric patients. These machines show us that new brain cells are born every day in specific brain regions. Recently developed scanners show us abnormal connections between neurons that are most likely secondary to gene mutations. We have identified genes involved in schizophrenia, autism, Alzheimer's, and other brain diseases and disorders. We know that brain diseases have a biological basis and are not just mental conditions that can be cured by a hand tap on the patient's shoulder. We need to go after genes and their twin brothers—proteins in their multiple shapes and functions. In the preceding pages I described for you how RNA and amino acids are involved in protein synthesis. You also know that proteins are a major component of every cell in your body. Proteins are chains of amino acids. Remember that many amino acids appear in nature but only twenty are used by our bodies. Equally important is remembering that shape determines function, and when proteins take the shape and function of an enzyme, they are responsible for many biochemical reactions. Proteins come in different shapes and sizes from just a few amino acids that form chains comprising hundreds of amino acids. I hope you can visualize how complex this protein synthesis is.

Let us imagine a protein composed of a thousand amino acids all packed tightly so it does not get entangled with other amino acids forming another protein or other intracellular subunits. It seems that the job that lies ahead of us is not a piece of cake. The beauty of the job that lies ahead of us is that we love challenges. Working on proteins that are particularly relevant to medical research and the solution of psychiatric diseases I have seen is an incentive I cannot fail to celebrate. You can help by attending seminars and workshops provided by researchers throughout our country.

Our Basic Carbon Composition

Recounting our story, you and I are basically made of carbon as a basic functional unit of multiple cells. Carbon, oxygen, nitrogen, and hydrogen compose about 95.6 percent of all the elements our body needs to survive. A carbon atom is a very dynamic atom. It makes many strong bonds with atoms and molecules of other elements that are necessary for all living things on our planet. Carbon atoms are part of amino acid molecules. All amino acid molecules must have a carboxyl group and an amino group, and both groups are joined to a carbon atom. The carbon atom that these two groups of molecules are joined to is known as alpha carbon. You now know that the basic unit of your entire body is protein. And proteins are long chains of peptides that in turn are made of amino acids. These protein chains coil up as if they are balls and are therefore known as globular proteins. So proteins are polymers of amino acids joined head-to-tail in a chain that could be several hundred units long. It is folded in a three-dimensional structure. The bond between two nearby amino acids is known as a peptide bond. The carbon atom is always involved in everything that is happening in your body. Just think about it; of the twenty types of amino acids found in all proteins, no matter which amino acid is chosen to form a polypeptide chain, it must have a carboxyl group at one end, known as the C-terminus, and an amino group at the other end, known as the N-terminus.

The beauty of all amino acids is that they repeat over and over again in all proteins. No matter what part of your body you begin to take apart for analysis, you will come across some of the same old amino acids. You may play around with amino acids adding growth or degrading factors, but you will end up with amino acids. The funny part of this game is that you will be learning about yourself. Even before you opened this book, the image of DNA and RNA likely came to your mind. These two acids—plus carbon atoms, proteins, and amino acids—are all about you. Later on, you became acquainted with the famous letters making up the two famous macromolecules of the two above-mentioned acids. The letters we are referring to are A, T, C, and G for DNA and A, U, C, and G for RNA. Please remember these different sets of letters for these molecules.

I have also used the word *nucleotide* quite loosely, but what is a nucleotide besides the usual connotation of a base? A nucleotide is in itself a molecule, which in turn is itself made of atoms. More carefully defined, a nucleotide is a molecule formed of nitrogen containing a ring compound that is joined to a five-carbon sugar. Basically, there are two types of nucleotides that researchers like to work on. Nucleotides that contain ribose are known as ribonucleotides (RNA). Nucleotides that contain deoxyribose are known as deoxyribonucleotides. These are long words that soon became shortened in daily conversation. In general terms, nucleotides are named after one of the bases they carry. That is why a thymine nucleotide is called thymine although it is implied that it carries one or more phosphate group. (Yes, this is the sugar-phosphate backbone that Watson, Crick, Wilkins, and Rosalind Franklin were struggling with while trying to come up with the double-helix structure.) Were the bases better off hanging outside the backbone or hanging inside, making hydrogen bonds? One nucleotide you will not be able to ignore while studying the two great nucleotides DNA and RNA is adenosine triphosphate, or ATP for short. It is an important energy-carrier molecule involved in many chemical reactions inside our bodies.

From Transcription to Proteins

Messenger RNA can leave the cell nucleus as a single strand and travel through the cytoplasm and land at a ribosome. This protein assembly factory also receives another single RNA strand known as transfer RNA. This molecule reads the message—three letters—that the messenger RNA clearly displays, and the ribosome pastes it together. After it is pasted together by the tiny globular ribosome, it is kicked out as a peptide. Multiple peptides form long chains that we know as proteins. Messenger RNA lands at a ribosome if another RNA molecule does not interfere along the way. Another RNA molecule may block messenger RNA from landing at a ribosome in the cytoplasm. In terms of space, time, and size, it is a long distance to go from one place to another. It is a long distance when we consider that it has to go through many smaller organelles, each of which is busy doing its assigned job. And there is, as mentioned previously, an RNA molecule, known as interference RNA, that bonds to target sites on a messenger RNA molecule, altering its shape and function. If we ponder the complexity of human life as a zygote grows into to a fully developed and healthy human being, we can see that the vast majority of us are blessed from the very beginning of conception.

It is not just interference RNA that may interfere with the formation of proteins; microRNA may do it too. There is a complete family of RNA that may interfere with protein synthesis. The first microRNA gene was discovered in 1993; five years later, RNA interference was discovered. The siRNA molecule made me grow goose bumps when I began to understand its potential power and intervention in protein synthesis and gene activation and suppression. Nicholas Wade wrote for the *New York Times* on June 21, 2005 that siRNA is "a system for silencing genes by tricking the cell into destroying the gene's messenger RNA before it can generate its protein product ... Together with transcription factors, micro RNAs may play a role in cell differentiation, the formation of many specialized types of human cell from a single generic type ... Micro RNAs create an environment, also tailored for cell type, in which some kinds of messenger RNA can flourish and others are diminished or repressed." Nicholas Wade used the word *tricks* in reference to siRNA's job.

I am trying to create for you a picture of the roles ribonucleic acids play in protein synthesis. Engineers had high expectations for their ability to make research tools and drugs from distinct forms of RNA molecules. However, RNA has proven a tough nut to crack, as some of the previously mentioned companies have found out. This is the oldest biologically-related acid we know of. When DNA showed up on Earth, RNA was an old man with gray hair. In the same issue of the *New York Times* listed above, Brian Libby wrote, "Dr. Kenton Gregory at St. Vincent Medical Center at Portland is one of a handful of researchers working to create replacement tissue from a naturally occurring protein, elastin." Elastin is known as a matrix protein because it holds the cells together to form tissues and provides natural support and flexibility. It gives skin and blood vessels their elasticity. Another matrix protein is collagen, which provides the tensile strength of tissues.

Proteins, Memories, and Eric Kandel

For many of us who are trying to understand ourselves, Eric K. Kandel's brilliant work and discovery of anatomical changes taking place at synaptic sites when permanent long-term memory is established in our brains is awesome and inspirational. To our delight, once more, protein synthesis takes first place with the complexity of transcription, translation, and peptide formation already discussed. Commenting on a discovery regarding drugs that inhibit the synthesis of proteins in the brain in addition to disrupting long-term memory if it is given during and shortly after learning, Louis Flexner, at the University of Pennsylvania, said short-term memory is not disrupted. This finding suggests that long-term memory storage requires the synthesis of new proteins. A very interesting and crucial question arose at that point: where does the new protein grow in a brain cell during the formation of long-term memory? Does it grow at the same site as short-term memory? Is long-term memory an outgrowth of short-term memory? Is there a specific region or cluster of neurons in the brain responsible for the formation of proteins for long-term memory?

Doctor E. Kandel has given us answers to our questions. His research at Columbia University provided him with a live tool he used wisely and fruitfully. He wrote, "Short term memory produces a change in the function of a synapse, strengthening or weakening preexisting connections; long- term memory requires anatomical changes. Repeated sensitization training causes neurons to grow new terminals giving rise to long term memory." Neuronal connections already exist; you activate them through a learning experience. Let's say you just learned the first three multiplication tables. You enjoyed it, and you repeated it several times. You even went to the next step of learning the next three multiplication tables. You felt great; you enjoyed the good feeling people experience when they accomplish something that satisfies them and the loved ones around them. You then went out and played ball, played with your dog, and even watched television for a short while. You could easily, without any effort, recall all the multiplication tables you had in your brain as long-term memory. This type of memory requires the synthesis of proteins that will be attached to, or seen as a protrusion coming out of, dendrites. It is as if a new small branch sprouts out of a neuron. This is what Dr. Eric. Kandell from Columbia University in New York City called anatomical changes in brain neurons.

A note of clarification: this is happening in neurons. The brain also has glial cells, but at the present, their role in memory formation is not clear yet. Dr. Kandel coninued teahing us by saying tht the plasticity of the nervous system (the ability of the nerve cells to change the strength and even the number of synapses) is the mechanism underlying learning and long term memory. He added, "One oh the most astonishing findings to emerge is that lomg-term memory gives rise anatomical chans in nerves cells". [37] Among long-term memories, there is one I have not forgotten since I first entered the physics' classroom: $E = MC^2$. Albert Einstein was such an outstanding figure in science that the professor at the University of Puerto Rico had his picture covering the main door as we entered his classroom.

Learning processes have been the subject of study, debate and speculation for millennia. Methods of teaching have been developed

---

[37] Eric K. Kandel, The Age of Insight, Random House (2012) p.309.

by many outstanding educators throughout the ages. Some teaching methods have survived the challenges of time, place, and the needs of society with slight revisions, while most lagged behind and disappeared. Determining how to pass on our body of knowledge—our wisdom—to the next generation is a great challenge to our society today. Equally challenging is determining the content of the body of knowledge we want our children and our children's children to learn in order to survive in a highly competitive society.

In the year 399 BCE in Athens, Greece, the greatest philosopher the Western world has known, Socrates, was forced to drink a preparation made from hemlock, a poisonous herb. He was accused of perverting the youth of Athens by unorthodox methods. The debate is even more challenging and controversial than ever before. We have issues like creation and evolution, science and humanities, and art and history. Some individuals are advocating reducing education in the last two areas to a minimum of time while raising science and technology to a top priority in school curricula. Is education a state responsibility, or should it rest primarily within the family? Should education of our children be left to educators and parents? Should politicians be in charge of our schools system just as they are in charge of legislating for all of us? The twentieth century left many scars on us inflicted by Nazis and Communist regimes. We are rapidly discovering how learning and memory take place in our brains. Sensitization and permanent neuronal structural changes can be easily manipulated in the classroom. These are hot issues that need our attention before it is too late.

More on Memories and Learning

Once I had grasped the idea of how learning and memories gain permanency in our brain cells, I could not stop searching. I turned to Dr. Kandel's engineered and exceptionally successful experiment on a manageable research toy, the aplysia. I was familiar with short-term memory, also known to many of us as working memory, and I was familiar with a patient who had surgery on both temporal lobes and had most of his hippocampi removed. After surgery, the patient, whom I will refer to as H. M, could not store memory for long-term use. All his memories after surgery lasted only a very short time. Somehow, functional memories could not find their way to other brain cells for storage and future use. H. M could not transfer his memories to a cluster of neurons for permanent storage. For example, if he were to finish a steak dinner and walk into a crowd on the street, and if you were to then ask him how he enjoyed his steak dinner, he would look surprised at your question and would simply reply, "What steak dinner are you talking about?" The steak dinner would be inside his stomach, but his brain would not have registered it. You may wonder if brain cells around the hippocampi in the temporal lobes could be generous enough under the plasticity theory and take over memory storage for H. M. It has been established that brain cells often take over functions not originally theirs. The physician who took care of H. M.'s health for many years reported that he never recovered his past memories after surgery.

Related to the hippocampus, Laura Beil, writing for *Science News*, said, "The hippocampus encodes and prepares new memories for storage, then dispatches them to different parts of the brain. In 1989 scientists reported evidence that the human hippocampus is not only a depot for memories, but also, a birthplace for neurons-thousands each month." The nursery for nerve cells is restricted to a raisin-sized region of the hippocampus called the dentate gyrus. Supporting the above statements, Fred Gage, a neuroscientist at the Salt Institute for Biological Studies in La Jolla, California, said that at any given time, about 3 to 5 percent of the cells in the dentate gyrus are in some stage of growth.[38] He further added that most of

---

[38] *Science News*, Vol. 179, no. 3 (January 29, 2011): 23–25.

the dentate gyrus is formed after birth; much of it is formed during the first four years of life. This is very important to know, as it relates to growth in children during a critical period for learning and forming new memories. Parents and educators—especially early childhood educators—should take advantage of this information to plan teaching lessons, as this relates to one of the main regions of the brain engaged in learning and memory formation. These are memories that will mold our personalities for life. For instance, the earlier you expose a child to other languages, the easier it is for him or her to learn them. The same is true for mathematics, science, astronomy, etc. Neuronal circuits and pathways, as well as synapses, are established and strengthened in preparation for adult life challenges.

Memories, the Brain, and Learning

Whether we like it or not, we are, for the moment, stuck with proteins. Therefore, I will continue the inquiry into proteins and memory formation. I carry toxic episodic memories in my brain that pop into my mind without an invitation. Likewise, I carry sweet and educational memories that make my life happy. My interest in protein synthesis, memories, and disease is not just an intellectual exercise; it is a very personal one. Besides inheriting thalassemia traits from my parents, I am shortsighted, and when I come out of a dark room and face light, I become almost totally blind for a few seconds. Of course I wear corrective lenses. I am not an albino person. I have been examined by an ophthalmologist, and he told me that I lack some vitamins and tissue in my eyes. The sun's rays blind me temporarily. My eyes are slow in adjusting from darkness to light. Some researchers have linked this to the pineal gland deep in the brain. This tiny cluster of cells is responsible for releasing melatonin and is also involved in the sleep-wake cycle. It seems that probing my brain is not a curiosity at all but a necessity discovery of myself through my strengths and liabilities.

I have repeated over and over again that the DNA molecule nucleus is the command center for everything that takes place in the cell. However, this seems to be only partially true. For example, strengthening or weakening a synapse is more like a localized issue than a command from the cell nucleus. For instance, if I do not repeat the order and names of planets in our solar system several times and read them aloud, the information will vanish from my brain. This is reflected in the popular phrase "Use it or lose it." Our brain has trillions of possible connections among dendrites. It can store many things, but it is my belief that the brain does not waste energy in trivial things that it does not use regularly.

The following might not be a good example, but I will give it to you anyhow. Coming from my home in the countryside, I rented a room where an elevated train passed by every half hour. The first nights were torturous; I filled my ears with strips of cotton in my attempts to go to sleep. Within six months, I did not notice the train passing by anymore. The cotton in my ears, the noise, and seemingly the train had gone forever. This is called habituation. My brain has survived thousands of years, and I hope it can survive many more. My brain uses synapses as needed; it also closes synapses not in use.

CREB Protein

I had read about protein kinase A traveling to the cell nucleus for the purpose of activating a gene or genes. In addition, I had learned from Paris that Jacob and Monod, the E. coli experts, had discovered that messages from a cell's surroundings can turn on genes' regulatory proteins. I assumed that a particular protein needed in a particular place was synthesized by the relevant gene or genes. However, in all honesty, E. coli has never been in my list of research models. Just to think that I did not invite it into my intestines to share my food and even create many unwanted symptoms in me makes me grow goose bumps. Perhaps it was my prejudice about these little parasites that prevented me from learning more about Jacob and Monod's work.

However, Watson's book on DNA and E. Kandel's book *In Search of Memory* made good use of E. coli secrets as revealed by those two researchers at the Pasteur Institute. E. Kandel came across a cyclic AMP response element-binding protein (CREB). CREB is a gene regulatory protein involved in many processes, including learning and memory. It binds to a promoter in a gene. In his laboratory, Kandel found that if he blocked the action of CREB in the nucleus of a sensory neuron, he prevented long-term, but not short-term, memory. This means to me that he was blocking signals to the central command center, the nucleus of the cell. Therefore, there were no orders for transcription to take place; consequently, there were no proteins in progression. Kandel wrote in his book, "Blocking this one regulatory protein blocked the entire process of long-term synaptic change!"

The activation of CREB leads to the expression of genes that change the functions and the structures of cells. CREB proteins come, in at least two forms—one for activation and another for suppression. This is the on/off switch nowadays. For me, it was not a surprise that protein synthesis takes place in the cell body. Oswald Steward from the University of California had shown us that protein synthesis also takes place at synapses. This was quite confusing for me because it reverses the whole functional process of a neuron. It bypasses electrogenesis, the hillock, and the neuron's nucleus itself. "The process initiates long-term synaptic facilitation by sending protein Kinase A to the nucleus to activate CREB, thereby turning on the effectors genes that encode the protein needed for the growth of new synaptic connections. The other process perpetuates memory storage by maintaining the newly grown synaptic terminals, a mechanism that requires local protein synthesis."[39]

I will add a final comment on CREB proteins and microRNA. Among microRNA, there is a miR-212 molecule, which is shaped like a hairpin. In experiments done with rats that were engineered to consume cocaine, the rats decreased their cocaine consumption when miR-212 levels increased. Perhaps this microRNA can be translated into a therapeutic tool for human addiction. Further experiments demonstrated that blocking the formation of miR-212 increased the appetite for the drug.

At present, we seem to have good command over RNA in its multiple forms. We can clone it, synthesize it and engineer it at will. This tiny RNA molecule volume is related to a CREB, a protein involved in long-term memory and anatomical changes in neurons. Eric Kandel tells us that protein kinase A goes to the nucleus of the neuron to activate CREB, which in turn turns on the gene that codifies the protein needed for growth of new synaptic connections. I beg you to bear with my curiosity regarding RNA. Whenever I turn a page in any book on biology and all its processes, there I find RNA involved in people's lives. I focus my eyes and attention on tools that may help ameliorate the pain of my brothers and sisters worldwide. The pain outweighs the benefit, if any, when cocaine or heroin hijacks the brain of a person. "Protection against cocaine addiction may be a side benefit of miR normal job of regulating CREB production and other biochemical processes in the brain," said neuroscientist Paul Kenny.[40]

---

[39] E. Kandel, In search of memory, p. 270
[40] *Science News, Vol* 178, no. 3 (July 31, 2010), 11.

Nobel Laureates Advocating RNA

THOMAS R. CECH, A 1989 Nobel laureate for the discovery of RNA's multiple functions and, in particular, ribozymes, wrote for Nobelprize.org in 2004 that he and Sidney Altman, independently of each other, found out that RNA could fold into complex shapes. Furthermore, it catalyzes biochemical reactions—a function previously thought to be restricted to enzymes. Thus, RNA was found to sometimes be an active participant in the chemistry of life, not just a passive messenger carrying molecules to a ribosome for protein synthesis. He added that RNA often controls the expression of genes—another role that had been thought to be at least mostly the domain of proteins called repressors and transcription factors.

The discovery of the ribosome as a ribonucleic acid molecule that assembles protein added scientific recognition of this acid's vital role in the chemistry, development, and evolution of life on Earth. It was seen as a home run for RNA advocates who were claiming that RNA had existed before DNA. RNA can replicate itself and multiply, while DNA needs RNA molecules for the synthesis of life's building blocks—proteins. One impressive feature of a ribosome is that it decodes a very large amount of messenger RNA to make protein for each organ, tissue, and function in the human body. Can you imagine multiple ribosomes putting together all the units from a messenger and a transfer RNA molecule? Basically, all orders come out of the cell's nucleus for the purpose of building a protein for a specific organ in your body. For instance, the blueprint for a specific cell in the prefrontal cortex, heart, liver, lungs, or a muscle in my left leg needs to be made in a precise fashion. There cannot be a mistake in the writing of this formula. This is by no means a small job even for a summa cum laude graduate. Remember: ribosomes are not located within the membrane of the nucleus to be supervised. Ribosomes are figuratively located miles away from the cell's nucleus, in the cytoplasm.

According to the above named researchers, a ribosome is composed of three RNA molecules (four in some species) and dozens of proteins. The most recent pictures of ribosomes at the atomic level show that they are, in fact, composed of RNA with no proteins in the vicinity. It is indisputable evidence they are indeed RNA catalysts. Ribosomes are in charge of directing mRNA and tRNA interactions in the process of protein assembly. At the same time that these discoveries were taking place, catalysts and enzymes were almost identical. In the past, RNA had been placed far behind, a passive agent not to be bothered with. However, research during the last decade of the twentieth century and the most recent discoveries have dramatically changed our concept about this very old ribonucleic acid and its multiple shapes and functions. During the birth of the double helix in the early 1950s, there were hardly any books on RNA—if any at all. They were hidden behind piles of illustrations of DNA and the double-helix structure.

The third millennium brought us a new experience with our grandpa, the RNA molecule. It turns gene expression on and off and plays a definite role in the making of proteins as well as chemical reactions within the cell. The beauty of this advancement in many areas of science is a blessing for humanity. The French and the American revolutions of the eighteenth century opened the door for the little genie that was kept imprisoned inside our skull to enjoy freedom. R. Descartes's "cogito ergo sum" could no longer be kept a prisoner of the pineal gland deep in the brain. The French philosophers of the late eighteenth century knocked down many walls, and the American brain built a new and democratic society. The human mind was set free. It could no longer be kept enslaved in the body by an antiquated and obsolete body of knowledge. We challenged authority handed down to us by the force of the sword, prejudice, and traditions that benefited and protected the few at the expense of the majority of humanity. Benjamin Franklin and Thomas Jefferson, although part of the American social elite, had minds that were not constrained by society's rules of behavior prevalent at the time. Franklin was a man of science ahead of his time. And we are still debating Jefferson's thoughts and actions. They were challenged by equally brilliant compatriots. We became the free men of planet Earth. Even the most repressive regimes used the word *democracy* in reference to our system of government.

A pioneer neuroscientist, Miguel L. Nicolelis, wrote that freeing the brain from the limits of our terrestrial bodies will allow the paralyzed to walk. Through this liberation of the human brain from the physical constraints imposed by the body, the disabled may rise from wheelchairs. Humans have shown that a brain can be directly linked to machines in a laboratory setting. He added, "A few years ago my group demonstrated the feasibility of linking living brain tissue to a variety of artificial tools. It will require a new generation of high density micro-electrodes that can be safely implanted in the human brain and provide reliable, long-term simultaneous recordings of electrical activity of thousands of neurons distributed across multiple brain locations." Doctor Nicolelis envisions bidirectional thought-driven interfaces operating myriad nanotools that will serve as our new eyes, ears and hands. [41] Similar techniques, such as deep brain

---

[41] *Scientific American* (February 2, 2011): 81–83.

stimulation (DBS), are used today to assist Parkinson's patients in gaining control of their bodies by, among other things, helping them control tremors and bodily balance.

Previously Unknown RNA Functions

Among RNA's previously unknown functions are the following: It can provoke termination of transmission of messenger RNA, it can interfere with ribosomes translating mRNA, or it can cleave mRNA into destruction. The RNA molecule has a cleavage mechanism that seems to work even in itself. RNA was studied alongside DNA for most of the second half of the twentieth century. However, with so many shapes and functions of RNA molecules having been recently discovered, it seems that, along with epigenetics, a new branch of molecular biology is needed. I am referring to RNA not only as a research tool but primarily as a medical drug treatment modality for many chronic and degenerative diseases that we are struggling to conquer. My concern and interest is stopping the suffering of millions of people around the world. RNA serving as a gene switch and as protein's most prominent assembly agent, in addition to its role as a breakdown facilitator, makes it seem likely that RNA will become an important player in medicine in the very near future.

Another attribute of RNA not previously suspected is its ability to come out as a double strand. Normally RNA is a single strand, meaning that only one copy is transcribed from DNA. However, given that it is a single strand with no partner to link with, it possesses the intrinsic attribute of folding back into itself. It can take any shape necessary for a function. There are RNA hairpins that are, functionally, double RNA strands. So a single strand of RNA can fold or coil itself into different shapes. For the sake of clarification, you can imagine it as a long spaghetti strand, a hairpin, a pair of pliers, a bead of a necklace or rosary bead, or a spherical structure.

It attaches itself to other molecules or allows other molecules to link with the histones in the chromatin. The linkage attribute takes many forms. In addition, there is the well-known RNA interference. This versatile molecule, RNA, can not only block protein synthesis but can also, with the assistance of enzymes like Dicer and Slicer, can cut double-strand RNA into twenty base pair fragments. One of these two strands can be transported to a matching sequence of messenger RNA for protein formation, while the other strand may face degradation, meaning destruction.

By now you are probably almost convinced about the crucial role of RNA in your life. But this short story about RNA does not end in here. It does not engage only in making proteins and interfering in gene function. Its future use depends on you and others taking your place among today's group of scientists.

And regarding meddling in genetic affairs, there is, of course, RNA interference. It has become a very important technology used by scientists to target gene inactivation in their never-ending search for drugs for the treatment of genetic-derived illness. Suppressing a gene may lead to the discovery of the causal factors of a disease or a syndrome affecting the brain or any other organ of the body. An RNA molecule that has the power to block the construction of the basic building parts of one's body during the translation process and can also inactivate genes is a powerful cluster of atoms. We need to understand and use the atoms that have formed into this molecule.

Please, bear with me, because I consider your brain as extremely important for us to ignore just a tiny bit of it. Your protein factory, the ribosome, consists of two subunits. It has a small subunit that decodes messenger RNA, and a second, larger, subunit that serves as a catalyst for peptide bond formation between the growing polypeptide chain and each new amino acid. "This is part of the process of translation I am committed to fully understand in eukaryotic ribosomes, which are much more complex and highly regulated than noneukaryotic ribosomes.[42]

A Little Bit of History

A little bit more history on RNA will not hurt anyone. Basically, over a decade ago, the race among scientists in the United States, Europe, and Japan insight into the synthesis of codons, amino acids, and proteins got researchers on both sides of the Atlantic busy at work in their respective laboratories. The RNA molecule's many roles and shapes have motivated many biologists to decipher this molecular acid's functions and shapes. The discoverers of the double helix in England were interested in this molecule, which appeared to be engaged in all aspects of our lives. Initially we were happy reading about and studying DNA double strands and RNA single strands—such as messenger RNA and transfer RNA, which are involved in protein synthesis. However, as more functions showed up, two young scientists: Craig Mello and Andy Fire, had not overlooked the power of the RNA molecule to intervene in the process of protein formation. Both men, working on E. coli, provided the first demonstration that double-stranded RNA could interfere with genes and even silence them. Do not forget this statement; it is very important.

Once again, the RNA molecule, which is so important during the synthesis of proteins, can intervene during the process of making proteins. Among other things, it can bind to a messenger RNA that comes from the DNA nucleus and derail its synthesis. In addition, it can also silence genes. Mello and Fire posed a huge challenge to biologists around the world. To think that an RNA molecule can alter and even stop the transcription process, a dominant function of the DNA molecule, was heretical.

---

[42] *Science, Vol.* 331 (February 11, 2011): 681.

Ever since I first read an article on this interfering molecule, I began to wonder what RNA had to gain when it decided to become DNA's workhorse. Within each cell, we have three wise and powerful macro-molecules working together to build, develop, and maintain a very complex organism like yours and mine. The three king molecules we are referring to are DNA, RNA, and Mitochondrial DNA Each one has a definite and clear role to play, and they have played those roles very wisely and beautifully. When I consider that silencing RNA molecules can be used to mark specific DNA sequences for deletion and guide modification of the chromatin structure, I get goose bumps. However, I try to bring relief to myself by telling myself that this molecule has worked remarkably well on me. The mitochondrion with its ATP molecule cannot be counted out in the balance of power among the three rulers of the eukaryotic cell. In a recently published scientific journal, a team of researchers who made use of mice linked mitochondrial DNA to health benefits. "In mice, a regime of endurance running for forty-five minutes three times a week over five months induced mitochondrial biogenesis, increased mitochondrial respiratory capacity, and prevented mitochondrial damage."[43]

An Established Fact—an RNA World

That an RNA world existed before a protein-coding one, which is the DNA world, is an established fact among modern scientists. The irony concerning this scientific truth is that it became accepted after the discovery of the double helix. Ironically, it provoked me into laughter and further self-questioning about RNA's functions. The seeming paradox that I am trying to deal with is that multicellular organisms developed after the protein-coding DNA world took over. However, the workhorse in protein synthesis is the RNA molecule

---

[43] Science, Vol. 331, March 4, 2011, P.115.

in its multiple forms. I previously stated that the DNA molecule incorporated RNA to do its bidding in the process of forming or constructing proteins. Similarly, I also pointed out that the wise DNA molecule incorporated the mitochondria. I have not been able to resolve this issue to my satisfaction. However, I am very happy with my progress in understanding both macromolecules and the sequencing of the human genome.

I would like to share with you some details from an excellent article written by Eric S. Lander in the journal *Nature*.[44] I will summarize just a few things that I found very interesting. The sequence of the human genome has dramatically accelerated biomedical research and our understanding of the biological functions encoded in the genome. The greatest impact of genomic sequencing has been the ability to investigate biological phenomena in a comprehensive, unbiased, hypothesis-free manner. In basic biology, it has reshaped our view of genome physiology, including the roles of protein-coding genes, noncoding RNA, and regulatory sequences. In medicine, genomics has provided the first systematic approach to discovering the genes and cellular pathways underlying disease. We have also identified locations where DNA and chromatin are modified, in addition to the study of inherited variations or somatic mutations.

Sequencing is also applied to RNA transcripts and RNA's multiple forms. Researchers have also clarified the human genome as consisting of about twenty-one thousand protein-coding genes. At the beginning of the genome sequence project, some researchers had put the number of protein-coding genes as high as one hundred fifty thousand. On the topic of small noncoding RNA, Eric S. Lander said that a few dozen miRNA have been shown to have key regulatory roles. Most recently, a new class of small RNAs called Piwi interacting RNAs has been discovered; they act to silence transposons in the germline. Transposons may be seen as drivers

---

[44] *Nature, Vol.* 409, February 15, 2001, p 860-69

of evolutionary innovations. In addition, we have benefited from sequencing the genome and identifying the locations and roles played by epigenetic markers. The vast majority of human variants have been discovered, and they are under intense analysis. In its clinical application, DNA sequencing is being increasingly used to assign patients with unclear diagnoses to a known disease. In psychiatric diseases, genomic studies have identified common variants in bipolar disorder and schizophrenia and rare deletions in autism.[45] I could not provide a complete account of the article, but I hope you can get access to the journal and enjoy it as I do.

Dr. Tom Misteli from the NCI and DNA

Ten years ago, a blueprint for a human being—the complete list of DNA's now famous letters, A, T, C, and G—was announced to the world. In celebrating a decade of genome research, Tom Misteli, a senior investigator from the National Cancer Institute in Bethesda, Maryland, said, "Biologists have known for long that DNA chromosome folds up in complex ways. They have now demonstrated that individual chromosomes occupy distinct territories in the nucleus and that some chromosomes prefer the nucleus periphery, whereas others like to cluster close to the core. Moreover, where chromosome resides, and which chromosome lie near one another, can strongly influence how cells function." (This author did not fail to notice the word *strongly* being used to qualify the influence on the cell). Tom further stated, "Where chromosomes lives seem to influence whether the genes it carries are turned on or off."[46]

A gene gets turned on after proteins known to us as transcription factors get together in regulatory regions of the gene. Our workhorse,

---

[45] *Nature* 470, no. 7333 (February 10, 2011) 187–97.
[46] *Scientific America* (January 2011) 70–73.

the RNA, this time in the form of protein RNA polymerase, comes in to transcribe the gene's DNA famous four letters. They are transcribed into the multiple RNA copies you are already familiar with. You will find DNA in the nucleus, while RNA is always moving around and changing its shapes and functions. Professor Misteli generalized based on his observations and said that researchers now know that the nuclear periphery has a silencing effects on genes, while the center promotes gene activation. He further elucidated the intrinsic working processes taking place during protein synthesis. He added that hundreds of genes that encode ribosomal RNA are transcribed together in the nucleolus—a nuclear substructure large enough that it can be seen under a microscope. [47] The nucleolus resides within the nucleus but has its own membrane. Do not look for it around ribosomes or nearby organelles.

Within the cell, there is traffic going on between areas of the cell. You will find organelles transporting subunits among the cell's basic components, from one station within the cell to another. You will also see polymers being brought into the cell through its membrane as well as sharing or transporting material outside to neighboring cells. If your organism has been invaded by unwelcomed bacteria and viruses, please, use your imagination for the traffic jam that will take place in millions upon millions of cells in your body. Your T and B cells, with legions of combatants, will order an attack to eliminate toxic invaders.

MicroRNA's Regulatory Role

The celebration of a decade of genome sequencing cannot ignore microRNA's regulatory role in gene expression. "Over the past decade, small RNA emerged as a new class of key regulators of eukaryotic biology. This diverse class of RNAs includes small

---

[47] *Scientific American* (February 2011): 71

interfering RNA, micro-RNA and PIWI interacting RNA, all of which associate with multiple protein components within a complex to regulate partially or perfectly complementary transcripts. Among the objectives the researchers were trying to understand is how mi RNAs bound to Argonate proteins, bringing about gene silencing. There is abundant literature indicating that translational inhibition and m RNA decay are coupled throughout biology." [48]

Unbelievably Naïve

My intention during this essay has been somewhat naive. My intention was to analyze and understand the code of life Watson and Crick began to play with during the middle of the twentieth century. I began with the double helix and proceeded to nucleotides, amino acids, and protein formation, which are among the most interesting things I found regarding our tree of life. I stopped for a while at the intriguing RNA molecule and the many curious and useful conformations and functions it engages in to serve us best. Of course I could not bypass epigenetics and multiple DNA mutations and histone modifications, which were briefly touched on. Perhaps my intention during this trip, although unconscious, is the very simple and naive goal of discovering my origin. I feel the nagging question of my original birth on planet Earth millions of years ago. I will not be able to answer it, but perhaps you will.

My own ancestors, including my grandparents, did not have this urge to explain the unknown. They relied upon generations of old tales that were often written in their sacred book, the Bible. More recently, our origin has been attributed to aliens or meteorites that came from another solar system. Both my ancestors and recent theorists are not too far apart; both place my origin outside planet Earth. However, for some reason, we tend to stick to the "chemical

---

[48] *Science* 331(February 4, 2011): 550–3.

soup" theory of Mother Earth. This wonder soup, under certain pressure and heat, supposedly attracted select atoms and molecules that in due time became living cells. It was a long, long time ago in our Earth's history that this spark of life took place, if that is indeed what happened. Most recently, science fiction and some respectable physicists and astronomers are advancing a theory that aliens with superintelligence may have visited Earth earlier in time and that we are just their guinea pigs, under constant scrutiny and observation. If this is the case, we look like animals under a laboratory's watch to see how we behave under certain conditions, how we take care of ourselves, and ultimately, how we take care of planet Earth.

Here and Now with Amino Acids and Proteins

All the above theories sound very interesting, but I prefer to stay close to Mother Earth. Instead of traveling out of our solar system in search of myself, I am stuck with amino acids, peptides, and Earth elements that formed my life here. Of course, the double helix and the genetic code are giant steps forward that hold my attention 24-7.

But when does life begin in a beautiful baby? A mother rightfully would say, "This is my baby, I gave birth to him." But did life begin at conception or during the first cell division? Perhaps it began three days later, during the morula phase of embryonic development, when there was an internal mass of eight cells? Someone might argue that we must wait another three days until the blastocyst phase takes place and cells begin their journey to specification. Someone else might say, "Life does not begin until cell specialization has begun and cells for a heart, a brain, lungs, a liver and the remaining organs and tissues of our body are in place." But now we are not now talking about the same thing. The original unknown was left far behind. But this is what happens during most of the meetings I have attended on this subject. An inquisitor may interrupt me and say,

"Oh, but this is democracy, my friend!" Of course it is democracy, but we are not talking about democracy. Besides, democracy in itself is in need of a new definition.

St. Augustine and Darwin Join Our Group

Suppose that our group of science-curious individuals is joined by someone like Socrates arguing in favor of his soul. A better scenario would be Aristotle, St. Augustine, Darwin, Einstein, and Oppenheimer joining our lively group. It seems to me that in this hypothetical case, we have a centrifugal chemical soup that could result in an implosion provoking a tentative truth. But whatever the implosion group may come up with, it is an educated guess of a group of people on a bit of dust in an infinite multiverse. Please do not blame me for beginning this mind-torturing dialogue. I blame the people I named above, including Watson and Crick. In the meantime, I will continue this journey without diverging any further.

In humans, brain growth and differentiation come under the control of genes around five to six weeks after conception. However, this prenatal phase of development is very critical for the fetus's health. The mother's diet and lifestyle, as well as environmental factors, will influence development of the fetus long after it leaves the comfort and protection of the mother's womb. Most of us are aware of the effects of cigarette smoking, drug abuse, biological predisposition, hunger, physical abuse, and chronic stress during pregnancy.

An estimated 100 billion neurons and around 800 billion glial cells need the correct amounts of vitamins and minerals to make their appropriate and corresponding shapes. They have to travel within the cell-building and discharging neurotransmitters necessary to achieve their final normal function. The effects of deficiencies in the basic elements necessary to build the amino acids that ultimately will form proteins for our body are well established around the world's scientific community. Following each step during fetal development is not the best thing to do with our present technology. We have our four-legged friend, the mouse, and other animals in our laboratories for our comfort and convenience. We use them to mimic human dietary deficiencies and possible health risks.

Stress and ADHD in Pregnant Women

Allow me to focus on stress as a subject of study. In a pregnant mouse, the exposure of the future mother's stress hormones can lead to anxious behavior and hyperactivity in the offspring. A relatively recent longitudinal study of over seven thousand mothers and babies (humans, not mice) at Imperial College London concluded that maternal stress may account for up to 15 percent of diagnoses of attention deficit hyperactivity disorder.[49] During the first decade of life, boys and girls spend much of their time playing together often under loving parental supervision and control. During this period of human development, the central nervous system is busy making many necessary circuits and pathways. During adolescence, sex hormones are having a big party inside your body. In the meantime, glial cells and neuron dendrites and spines are trimmed to serve specific purposes corresponding to biological, physiological, and behavioral stages of development. You have association neurons; motor neurons; visual, acoustic, and limbic systems; and many other brain cells making appropriate adjustments to best serve you.

The prefrontal cortex, the brain area that controls cognition and high-level decision making is the last neocortex region to complete maturation. Unfortunately, this occurs during the time when adolescents feel that they have to experiment with everything around

---

[49] *New Scientist* (April 4, 2009): 28

them that adult people take for granted. They experiment with drinking alcohol, smoking, and driving. The likelihood of being in an injurious or fatal automobile collision is higher during this period of a human's life than at any other period of time. Unfortunately, this takes place while our brain's decision-making cluster of cells is still in the process of making corresponding axonal connections. This does not mean that connections are permanently closed. Synapses are busy receiving and sending chemical messages with potential new pathways. Adding to adolescent turmoil, a dreadful brain disease, schizophrenia, takes its toll during the late teens and early twenties. Brain areas like the nucleus accumbens and ventral tegmental area are easily recruited by our reward system, which ends up hijacking the whole brain; consequently, addiction takes over the owner's life. According to reliable studies in various laboratories, once an addictive drug has hijacked the brain, it makes permanent changes to those areas. The addicted individual exists to sustain an addiction that has taken over part of his brain that dictates his behavior at all costs. The addicted individual does not enjoy his drug, because the state of feeling that you and I experience is no longer part of his brain's behavior.

Daydreaming

Our hypothetical group in which St. Augustine and Darwin joined us to help us come to an educated guess is based on many TV documentaries about people who have attempted to solve our problems. The groups have ended, in the majority of cases, just as they were in the beginning—with each member holding to his or her beliefs. We were not dreaming, were we? I suppose we were daydreaming for a short time. But is daydreaming a resting period for the brain?

Daydreaming has had multiple interpretations and meanings for

DNA-RNA Research for Health and Happiness

many people around the world. Some researchers have concluded that it allows brain memories to be analyzed and organized into content-related compartments in brain cells. Occasionally, researchers and psychologists have associated daydreaming with neuron affinity, circuit and pathway compatibility, and normal functioning. Much less reliable are philosophical and religiously inclined individuals who postulate that daydreaming is the time for the soul to leave the body momentarily. They maintain that the soul leaves the body to communicate with its superiors, whatever those might be. All of the above persons defend each other's positions with about equal tenacity and conviction. Unsurprisingly, when I began to read an article on daydreaming that related to brain diseases like schizophrenia and autism, my brain was half asleep. Occasionally, for some individuals, daydreaming suddenly lights up the brain like the sun at noontime. Daydreaming can be a very sweet thing—especially when boys dream of a girl that exists in dreams only. Excessive daydreaming can be an escape from reality and the challenges of everyday life. But many poets, storytellers, and novel writers claim that they produce their best literary work while daydreaming.

The Brain—Just 2 Percent

The brain, the daydreaming part of your body, is on average only 2 percent of your body mass, but that brain of yours consumes 20 percent of the calories you eat. I could not help it; my mind just immediately jumped to implicate the replication, transcription, and translation processes RNA is involved with to form proteins in my body. The mitochondria and its energy-producing molecule, ATP, popped up in my thinking processes and alerted me not to count them out. The truth is that the brain, while seemingly resting during sleep—especially REM sleep—is not resting at all. *REM* stands for rapid eye movement. During my REM sleep, in all probability, my brain consumes energy at a higher rate than while I write these lines. Also, your brain does not store energy for leaner times. If calories are needed and you consume no food, fat that has been stored in your belly, buttocks, thighs, and elsewhere will go to feed your brain. If for some reason you risk resisting your brain's demand for calories, you will be on the losing side of the battle. Your brain has a mechanism that will convert whatever you have stored in your body into calories until there is no more to devour and death arrives.

The scenario I presented you with regarding a brain's need for a huge amount of calories is not a hypothetical case. A group of researchers with a curiosity in the brain's need for calories wanted to know why and how the brain consumes so much energy. It consumes energy even during the resting state of brain functioning. To tackle this issue, they have used the most modern technology at their disposal—PET and MRI scanners. During one experiment, brain cells were fully activated during rest time but quiet—seemingly resting—as soon as the subject began to mentally work on an exercise. This is seemingly paradoxical, but it was a very interesting experiment. It seems that if I work on solving a puzzle, my brain cells leave me alone and go about their own business. However, if I go into a resting state, the brain's neurons go to work. My brain seems to have a funny work habit.

Raichle, Shulman and the Default Network Mode

Gordon Shulman, a researcher, went through 134 brain scans and found that regardless of whether the subjects in the experiment were reading or watching shapes on a screen, the same group of brain cells showed decreased activity. They grew dim as soon as the subjects began mental concentration. The researcher called this the default mode. Does this mean that brain cells chat among themselves when not busy doing routine life chores and complex intellectual work?

This is even more curious. During the resting or chatting time, parts of the network consumed 30 percent more calories, gram for gram, than nearly any other part of the brain.[50] You could argue that the brain's multiple neuronal processes, including electrogenesis, memory formation, and storage, and maintenance of the autonomic nervous system, demand caloric energy in vast amounts. Our brain is always busy daydreaming, solving routine daily life chores, planning self-protection or trying to understand Einstein's quantum theory. That brain of yours deserves a lot of respect. It led you out of the jungle and is planning to take you away from planet Earth and to where no man has gone before.

These researchers used very modern tools to come to very interesting conclusions I cannot avoid sharing with you. Marcus Raichle says that there is a huge amount of activity in the resting brain that is unaccounted for. Making use of PET scanning, he noticed that some brain areas seemed to go full tilt during rest but quieted down as soon as the person exercised. Neurons chatter nonstop to one another when an individual is unoccupied, but as soon as a person begins a task requiring focused attention, the chattering quiets down or stops. I became even more interested in this discovery when I learned that the median prefrontal cortex and the hippocampi become involved in this default network. You already know that the hippocampus is involved in the formation and retrieval of memories. The prefrontal cortex is highly engaged in decision making after receiving input from many important brain areas. A possible task for the default network would be linking multiple neuronal circuits and pathways among clusters of brain cells and providing the brain with an inner rehearsal mechanism for considering future actions and choices. Using fMRI, Malia Mason of Dartmouth College in New Hampshire equated this network mode with daydreaming. Randy Buckner and Daniel Gilbert from Harvard University "see it as the

---

[50] *New Scientist* (November 8–14, 2008): 29.

ultimate tool for incorporating lessons learned in the past into our plans for the future." Marcus Raichle now believes that the default network is involved in selectively storing and updating memories based on their importance from a personal perspective. He added that rather than burning the extra glucose for energy, it is used as a raw material for making the amino acids and neurotransmitters it needs to build and maintain synapses. [51]

Dr. Marcus Raichle's Opinion

However, another scientist using MRI reported that daydreaming took place when the default network was active. Default mode was coined for an unoccupied brain. The brain is not working on your own business. It must mean that you are consciously working on something. Marcus Raichle from Washington University at St. Louis believes that the default network is involved selectively storing and updating memories. Another discovery of Reichle on our brain is that it has a sweet tooth. It devours a huge amount of glucose that is far out of proportion to the amount of oxygen it uses. He believes the extra glucose the brain consumes is used as raw material for the making of amino acids and neurotransmitters that the brain needs to function properly. Now comes the big surprise: Raichle and two more researchers found that the default network's pattern activity is disrupted in patients with Alzheimer's disease. Most Alzheimer's patients' brains are not occupied with anything, because their memories are gone, and their neurons are blocked by plaques. And organelles inside their neurons are entangled with one another, and each one is unable to send or receive messages.

The point I intend to make is that the presence of chattering among brain cells in people suffering from advanced Alzheimer's disease is very questionable. However, Raichle's findings indicated

---

[51] *New Scientist* (November 8–14, 2006): 31.

that the default network also turns out to be disrupted in other brain maladies, including depression, attention deficit hyperactivity disorder, autism, and schizophrenia.[52] Neuroscientist Elkhonon Goldberg, from New York University, commented on the frontal lobe and Alzheimer's disease: "My collegians and I have shown that frontal lobes become dysfunctional at a very early stage of Alzheimer's type dementia ... The frontal lobes are more vulnerable and are affected in a broader range of brain disorders: neuro-developmental, neuro-psychiatric, and so on, than any other part of the brain."[53]

The findings in this article made me scratch my head, because the brain in subjects with schizophrenia and ADHD—and autism, to a lesser degree—is in high gear most of the time. Brain activity in patients diagnosed with major depression, as shown in scanners, seems to be taking a siesta time most of the time. The way I see it, brain neurons in schizophrenics and sufferers of ADHD are chatting most of the time, forming hallucinations and delusions. On the other hand, we have the deeply depressed subject who complains that she or he has nothing to report during psychotherapeutic sessions. Many depressed patients complain that their brains seem to be empty. I recall one saying to me, "My head is like an empty shell; nothing goes in, and nothing comes out." The medication prescribed to them is intended to activate their neurons.

Pete, the Autistic Boy in my Office

My experience with autistic children has been limited, but I discovered that they enjoy themselves when I gave a patient of mine, Pete, a game or puzzle to work on. I sat quietly next to him. I remember Pete telling me the kind of puzzle he used to love to

---

[52] *New Scientist* (November 8, 2008): 31.
[53] Elkhonon Goldberg, *The Executive Brain* (Oxford: Oxford University Press, 2001), 115.

work on. He strongly disliked noise and excessive sunlight. He used to pull down the shades in my office. Another interesting thing about his behavior is that he did not like his brother or his mother to share our sessions. During most of the time we spent together, he would begin our conversations. At the end of each session, Pete would often ask me for an object to take home. I liked him asking for something to take home from me. It had significant meaning for me. It took me time to refer him to a licensed trained therapist. I had neither the training nor any understanding of the neurology underlying autistic behavior. I interpret him taking home toys from my office as taking home part of me, thus bending isolation through trust. The key question I was always trying to deal with while in practice was how to use brain plasticity to normalize brain circuits and chemistry. With Pete, I confess I was frustrated and sad I did not know how to help him.

Autistic children are hypersensitive to sound and touch. For them, soothing birdsong sounds like a drill sergeant during basic training, or an eighteen-wheeler blowing its horn on the highway. Most recent scan studies have revealed that autistic children process sound in an abnormal fashion. Likewise, a soft touch on the skin of an autistic child seems to the child like a rough game of soccer or football. "Autism is largely an inherited condition. If one identical twin is autistic, there is an 80 to 90 percent chance the other twin will be autistic as well." [54] Autistic children do not seem to be able to discriminate between degrees of stimuli; consequently, they have trouble tolerating stimuli. It seems to me that the autistic brain closes up neuronal circuits and pathways in selective brain areas ahead of time, before it can establish stimulus control levels. They are unable to decrease the force of the sound of thunder and bring it down to a tolerable level.

---

[54] Norman Doidge, *The Brain That Changes Itself* (New York: Penguin Books, 2007), 77.

Special training and teaching in general will help to reeducate the neuronal circuits and pathways in autistic individuals' brains, leading to normal responses. Autism is much more prevalent in boys than girls. The autistic child is not avoiding or rejecting anyone; he or she is protecting himself or herself from experienced harmful stimuli. The earlier we establish the diagnosis, the sooner we can begin synaptic reeducation.

Isolation and Loneliness in Schizophrenic and Autistic Individuals

Schizophrenic and autistic individuals tend to be socially isolated people. Their brains seem to be on the alert most of the time. It seems as if they expect rejection from all sources. Consequently, they respond by escaping into their own shell, which they find comforting and safe. Researchers have found that the amygdalae and the prefrontal cortex are more active than most other brain regions. We may also interpret the hyperactive amygdalae and prefrontal cortex as an adaptation or adjustment of those areas. This appears to be autistic children's behavioral response toward people they come in contact with. In autistic and schizophrenic individuals, there may be a biological predisposition that is triggered or activated by outside behavior. Their brains do not engage in regular daily problems and conversations, and consequently, they turn inward. I cannot call them daydreamers, because the brain is very active in forming or recalling hallucinations and delusions or forming response designs. You will find similar brain behavior in manic-phase bipolar individuals. The prefrontal cortex is actively judging rejection signals; therefore, commonsensical behavior, decision making, and academic accomplishment become deficient. Consequently, focusing attention is a major problem for these individuals. This may lead to feelings of chronic sadness—not that sadness is inherited, but along with loneliness, it is an undesirable behavior that stems from a biological predisposition. The environment during the early years of development is particularly critical.

We must remember that the brain is not a rigidly wired organ of our body. New neurons form in the brain every day, particularly in the hippocampi and ventricles. In addition, we have synapses that connect neurons and establish new pathways, reinforcing good and appropriate behavior and thinking. Brain plasticity exists for us to take advantage of and put to work with people in need. The brain learns new things every day. It learns something new, and the region or cluster of neurons responsible for processing sound and touch in autistic children have to adapt to new levels of signals.

Neurobiologist Jeff Lichtman, from Harvard University, hopes that studying the structural changes in the brain will eventually show him how information is encoded in our brain wiring. Using a method he dubbed *Brainbow*, a small piece of DNA with genes that code for random amounts of yellow, blue, and red fluorescent proteins in nerve cells are inserted into mice. Through this method, his team can give individual neurons different colors. Those colors will allow researchers the opportunity to follow and determine synapse and axon connectivity. Connections offer a way to investigate thought disorders, such as mental illness and learning disabilities. Further addressing the wiring or connections between neurons, he stated, "The plasticity of the brain's wiring is one of its greatest strength and scientists hoping to learn more about how we learn and adapt as we grow."[55]

---

[55] *Science Illustrated* (March–April 2010): 35.

Your Brain and Plasticity

ARISTOTLE, THE ANCIENT GREEK philosopher and tutor to Alexander the Great, proclaimed some of his findings based on personal observations. They became absolute dogmas for almost two thousand years. Accordingly, the sun and stars were thought to moving around planet Earth as if Earth were the center of the universe. Our dear and sweet Mother Earth was believed to be a flat mass of land until a few hundred years ago, when Christopher Columbus helped tear down that fallacy. People become dogmatic and stick to beliefs and practice even when facts contradict them. The Earth was believed to be the center of the universe. It fit well with Christian belief and practice. Christian teachings hold that Christ came to Earth to save us from eternal death. Men and women- especially Christians- on this planet inherited this pardise of ours by faith and believe in Christ. Therefore, men and planet Earth became the center of the universe. Christian theology on west and eastern Europe held Aristotes`s "truth" in hgh esteem. A very small group of individuals questioned both, Aristotles and Christian held dogmas. Some men were encarcerated and killed if they did not recant.

And I should point out that this does not happen only to the uneducated masses. The theory of localization as pronounced by Paul Broca and Carl Wernicke became an axiom, an unchallenged truth, that still persists in some professional quarters. Dr. Broca had a patient that answered all his questions with the word *tan* after suffering from a stroke in 1861. After the patient's death, Dr. Broca performed brain surgery and found out that an area in the left hemisphere, now known as Broca's area, was gone.

A few years later, Dr. Carl Wernicke had a patient with a similar problem, except Wernicke's patient spoke in an unintelligible fashion. It sounded more like gibberish than a normal conversation. It lacked conjunctions among words and phrases. After surgery, a region in the left upper temporal lobe, now known as Wernicke's area, was kaput; it was damaged. Based on these two cases, followed by many well-attended professional conferences, the one function, one location theory was established. Both areas, Broca's and Wernicke's, are located in the left hemisphere of the brain; thus, another "truth" was established, and the left hemisphere was deemed dominant regarding language.

A Nondogmatic Believer, J. Cotard

However, in 1868, a nondogmatic believer and science-oriented physician, Jules Cotard, studied children who had suffered serious early disease. He found that those among the children in whom the left hemisphere, including Broca's area, was wasted away, could speak normally. In 1876, Otto Soltman removed the motor cortexes from infant dogs and rabbits. The motor cortex is responsible for movement, but surprisingly, Soltman found out that his little rabbits and dogs could move with their motor cortexes gone. Cotard and Soltman had not only disproved the one function, one location theory but, more importantly, had discovered the plasticity of the brain. The experiments they performed showed that the brain was plastic enough to reorganize itself to meet demands of daily life chores.

From Broca's surgery in 1861 until now, 150 years have passed, and we still have many professionals doubting the brain's capacity to take over functions damaged by strokes or disease. Furthermore, despite massive evidence using the most modern scanners, it is very hard for some individuals to accept the growth of new neurons in the brain. Plasticity of the brain must be understood as not only taking place at the synaptic level; it is also possible for clusters of neurons to take over functions from one brain hemisphere to another.

To revisit brain neurogenesis, a recent study reported that the subventrical zone around the anterior lateral ventricles in infant humans is highly active, producing many tangential migrating immature neurons. However, beyond eighteen months of age, both proliferative activity and cells expressing markers of immature neurons are largely depleted.[56] Arellano and Rakic, commenting on the above quotation, wrote that Sanai and others showed that the migratory cell streams departing from the subventricular zone (SVZ) are present in human newborns but disappear by the eighteenth month after birth.[57]

1876 Was a Long Time Ago

In reference to Otto Soltman's work with dog and rabbit motor cortexes, we do not know exactly what he did. The motor cortex is a complex system composed of multiple regions in the brain. His surgery was performed on infant animals when the wiring of their brains was taking place or was in the process of consolidating functions. It is unknown to us which areas or groups of motor neurons he removed. It was 1876, and not a single machine we have today to slice and examine brain organs, tissues, and functions existed at the time.

I decided to write a few observations on this subject so you

---

[56] *Nature* (October 20, 2011): 382–5.
[57] Ibid., 333.

can appreciate the effort a Parkinson's disease–affected person goes through to be able to get up, move, and walk. Similarly, many family members, friends, and valiant soldiers come home with spinal and brain injuries. The motor cortex has multiple loops that facilitate lateral, horizontal, ascending, and descending connections with the neocortex and spinal cord, among other things. It is divided into three reciprocally interconnected areas: the primary cortex, the supplementary motor area, and the premotor area. In the primary motor area, neurons correlate with a variety of movements and the firing of individual cells. The secondary motor cortex is involved in planning movements, and the premotor area is particularly concerned with planning movements that require sensory cues.[58] For Parkinson's patients and combat soldiers with central nervous system injuries, it is very painful to overcome the limitations imposed by these brain deficits. Both physical and psychological stress are coupled with these issues. These are adult men and women who have to learn anew how to get up, move, and walk. They have to force neurons in different brain locations to learn to do a job they have not done before. It is a physiological and psychological demand on the patient to force brain cells to take on new functions. I get a feeling I cannot share with you in words when I see patients, family members, and friends walking and doing things they could not do before.

P. Bach-y-Rita and Plasticity

In 1969, Europe's most prestigious scientific journal, *Nature*, published an article that made many people shake their heads in disbelief. The article, whose leading author was scientist and physician Paul Bach-y-Rita, described a device that enabled people who had been blind from birth to see. My initial response would

---

[58] A. Longstaff, *Neuroscience* (Milton Park, England: Bios Publications, 2000), 221–2.

be that he was talking about science fiction or the biblical story of Jesus of Nazareth two thousand years ago; however, it has been proven that it was neither. Almost half a century ago, Doctor Rita had rejected the above dogma of one function, one location and proclaimed that our senses have an unexpectedly plastic nature. Bach-y-Rita maintained that if one location is damaged, another can sometimes take over for it through a process he called sensory substitution. He claimed we see with our brains, not with our eyes. This reminds me of Geordi La Forge in Captain Picard's starship *Enterprise*. Dr. Rita developed ways of triggering sensory substitution and devices that give us super senses. He laid the groundwork for the greatest hope for the blind: retinal implants, which can be surgically inserted into the eye.[59]

Self-Isolation, Not Loneliness

Self-isolation was widely practiced by monks during the early years of Christianity. Socrates's dual nature of humanity, body and soul, was well accepted by Christians for the purpose of restraining bodily imperfections and desires. Consequently, it advances the soul's journey while one is on planet Earth. Today we advance the needs of the body and the thinking processes of our brains. The soul has been left to follow its own destiny. Self-imposed isolation, except during hibernation in some animals, is not the preferred way of life for any living being.

However, my curiosity regarding all living things came to a sharp focus on a bug that had been buried in ice for one hundred twenty thousand years. Do not hesitate; there is nothing wrong with your eyes. I wrote one hundred twenty thousand years. Scientists coaxed the bug back to life. The bug had been lying dormant three kilometers deep in the Greenland ice sheet. I asked myself what it ate

---

[59] Doidge, *The Brain That Changes Itself*, 10–17.

to survive so many years in such an inhospitable environment. The research team leader, Jennifer L. Curtze, listed dust, bacterial cells, fungal spores, minerals, and other organic debris as food sources. She further added that the best medium in which to preserve amino acids, organic compounds, and cells is ice.[60] I would like to add that Jupiter's moon Europa and both poles on planet Mars may well be places that primitive life forms thrive on. In 2010, scientists from NASA reported that a bacterium in a lake somewhere in California could survive in arsenic acid, which is fatal to humans. This is not science fiction or a matter of faith; these are scientific facts.

The journal *New Scientist* ran an article with the headline "Goldmine Bug DNA May Be Key to Alien Life." The article claimed that "the organism's ability to live in complete isolation from other species, or even light or oxygen, suggest it could be the key to life on other planets."[61] This strange bug was found in a South African gold mine nearly three kilometers beneath Earth's surface. Dylan Chivian, from Lawrence Berkeley National Laboratory in California, analyzed the newly found bug and said that the bacterium gets its energy from the radioactive decay of uranium in the surrounding rocks. It has genes to extract carbon and nitrogen from the environment, both of which are essential for protein synthesis.

Evolution, adaptation, and surprises of all sorts continue to keep my brain in a permanent state of alert for new discoveries. In an October 2007 edition of *New Scientist*, I read that there is a possum that slept for a year. The researcher claims that a possum he was experimenting with, after stuffing itself with food, curled up and went to sleep in his laboratory for a full 367 days. Another animal, *Zapus princeps*, had slept for 320 days; however, an Australian pygmy

---

[60] *New Scientist* (June 20, 2009): 8.
[61] *New Scientist* (October 18, 2008): 17.

possum broke the record by using one-fortieth of the energy it does while awake.

The Hunt for Diseases and Genome Sequencing

The hunt for diseases in genes has engaged many outstanding scientists around the world. Initially the main focus of attention was centralized in DNA's mutations, which were suspected in fatal diseases such as multiple sclerosis, Alzheimer's dementia, schizophrenia, and autism, among others. In some of these cases, the gene may appear normal, but there might be a surplus or deficit of DNA sequences. Among copy number variations, chromosome 21 can have three copies instead of the normal pair, consequently provoking Down syndrome. Past literature refers to children with this syndrome as being Mongolian-looking. This type of rare variation was considered conducive to diseases in most cases. However, further studies revealed that variations in gene quantity is not an anomaly but on the contrary is quite common.

In 2006, a group of geneticists analyzed DNA from 270 individuals and identified an average of forty-seven copy number variants per person. The genome sequence of C. Venter, a pioneer in the field, was found to have sixty-two copy number variants. As already mentioned in another chapter, J. D. Watson's genome also revealed many copy number variants. These two outstanding scientists are doing exceptionally good work for humanity, so one cannot help but wonder what the role or function of those rare variants is. However, "copy number variants have been linked to disease like autism and schizophrenia."[62] It can be argued that Down syndrome's rare variants should not be placed alongside autism and schizophrenia, and I fully agree with this. Autism, schizophrenia, and bipolar disorder are considered them syndromes. Their causes are not

---

[62] *Scientific American* (June 2009): 24–25.

only a matter of genes, proteins, cell signaling, neurotransmitters, and glial cells but also myriad other causal factors.

By the way, copy number variations have been described as alterations of the DNA of a genome. Such variations provoke the cell to have an abnormal number of copies of one or more segments of the DNA molecule. The alteration may be an increase or deletion of a segment of the genome. In some cases, the variations may each be as much as 10 percent of the human genome. They may pose possible serious risk factors for diseases or abnormalities nobody wants to have. Scientific opinion on the etiology of these genome variations seems to be divided at present time. There is evidence that they are inherited, while there is also scientific support for de novo mutation, meaning that it may be seen as spontaneous mutation.

Mobility within the Genome

It is important to note that the genome is not a long strand of beads tied to each other without mobility. The genome suffers deletions, modifications (as in SNPs), duplications, and multiple interventions by genome fragments. There are several gene segments that are involved in the multiple processes carried out by DNA and RNA. Consequently, variations in general seem to be the rule instead of the exception. Copy number variations may take place in a single gene or be extended to a neighboring set of genes. Variations in the quantity of genes may be responsible for a substantial amount of human phenotype variability. Besides, we have to consider behavioral traits and disease-provoking sensitivity secondary to rare variants.

James Lupski, a clinical geneticist at Baylor College of Medicine in Houston, Texas, stated online in the prestigious scientific journal *Nature*, "Highlighting the importance of rare genetic variants in causing diseases … even one copy of a certain gene can have profound consequences for brain development and mental disabilities." I must ask, how about gene mutations that do not code for proteins and become degenerated debris in a cell? Mitochondrial DNA is also subject to multiple modifications and problems. This female contribution to our body and heritage can be a blessing or a liability. Dominant and recessive variants on both sides can contribute to make me a genius or a disabled individual unable to care for myself. One of my parents gave me an unwanted present—thalassemia minor. Purification of our genome has to be kept on the table for future laboratory consideration and analysis.

Reinforcing Erroneous Assumptions

Not long ago, many scientists believed that silent mutations were inconsequential to health because those changes in DNA would not interfere with protein synthesis. Although some disorders were traced back to a silent mutation, the general consensus was that the culprit could not be a silent mutation. It was argued that this would go against the general observation of most researchers in the area of molecular biology. Reinforcing that erroneous assumption was the discovery that many silent mutations in various species were preserved over a long period of time. In addition, the consequential changes in many species made protein production more efficient. However, there was a pronounced and significant exception; it did not apply or did not work in humans.

As human genome sequencing technology advanced, scientists began to question their erroneous assumption. It was an assumption based on scientific observation and logical conclusions. By now you are most likely convinced of how efficient processes of protein synthesis have developed in our bodies. For instance, you may recall that in RNA chains, the letter U, for uracil, is substituted for the letter T, thymine, in DNA chains. During this complex process of protein manufacturing, there is a nucleotide change for the benefit of life in general. Simply reasoning it out, the information encoded in nucleotides is converted into the language that builds amino acids, which in turn will finish up as proteins. During the whole process of manufacturing proteins, there is a set of rules and steps that should not be broken. This set of rules governing DNA and RNA is widely known as the genetic code. Among the rules is the formation of messenger RNA and its editing process (splicing noncoding DNA), and transfer RNA, which bears amino acids to their destination—the ribosome.

The job is not as easy as many people think it is. Take, for instance, the amino acid leucine. It has six codons, all of them coding for the same amino acid. The codons are UUA, UUG, CUA, CUC, CUG, and CUU. On the other hand, there is the amino acid tryptophan, which has only one codon—UGG. The amino acid serine has six codons, and tyrosine has two codons. In all, there are sixty-one codons codifying for the twenty amino acids. In addition, there are three stop codons. As I said earlier, this set of rules cannot be broken or altered; otherwise, we will end up with multiple health problems and diseases. Do not be frightened; I already told you that the cell possesses a self-correcting system to ameliorate problems.

Just a few more observations on this subject may be helpful to some students who would like to go a little further into gene mutations and related issues in biology. A single-letter change to the code we have talked about, which ends up as a healthy and efficient protein, is known among neuroscientists as a point mutation. Such a mutation can provoke a changed codon to incorrectly form an amino acid. In professional language, this incorrect formation is known as nonsense mutation. Consequently, the resulting product would be a shortened protein. Therefore, single-letter changes in the chains I am writing about can cause several problems. Among those problems are the incorrect formation of a stop codon, which provokes a change in an amino acid, culminating in a longer or different protein. No wonder scientists in the field of molecular biology could not point with certainty the loci of problems and come out with correct conclusions.

Scientists began to challenge the prevailing theory on silent mutations about thirty years ago. By the middle of the 1980s, scientists began to realize and accept that silent mutation could intervene and improve protein synthesis, at least in single-celled organisms. With new and improved research technology, scientists began to discover that cells show preference for some codons, meaning that cells do not treat codons on an equal basis. It seems that cells employ protein synthesis efficiency selectively. Just as we tend to care for and breed animals and birds that satisfy our curiosity, needs, and interests, mammalian genes tend to favor certain codons over other codons.

The genetic code that I have referred to many times is nothing more than a set of rules in a cell. This set of rules dictates how information encoded in genetic material, such as DNA or RNA sequences, will turn out as amino acids and ultimately come out as proteins from a ribosome. The in-between processes of nucleotide formation, triplet formation, transcription, and translation are indispensable rules that cannot be avoided or broken. These are in-between steps the genetic code has imposed on the process of building proteins before the final product will be delivered to specific organs within your body. You owe it to yourself to learn how you are constantly rebuilding yourself to keep your body beautiful, handsome, healthy, and intelligent.

A Bug inside My Brain?

IT SEEMS THAT IN this essay I have picked on the translation process as the villain responsible for many of our diseases and disorders. Please believe me when I tell you I am not prejudiced against translators or anyone else. On the contrary, I recognize how difficult it is to take an abstract concept and decode it into everyday parlance. Sequencing the human genome and learning about all the debris, markers, mutations, modifications, interventions by short segments of genes, etc. makes me admire geneticists with awesome respect. Discovering and translating short nucleotide sequences responsible for flagging or alerting boundaries of an exon is awesome and worthy of praise by all of us. This editing process becomes even more interesting when one finds SNPs engaged in or influencing this already complex process of protein manufacture. During this scenario, there is no room to blame translators or researchers; it is caused by the genetic code itself as expressed in DNA and RNA.

I have said several times that the brain matter inside your skull is basically composed of glial cells. Some experts in the field claim that over 90 percent of your brain matter is not involved in making good and useful decisions for you. Few scientists who like to speculate on this subject claim that around 93 percent of our bodily flesh—but in particular, our brains—is a carryover from the development of the earliest organism. The human brain can be seen as a huge mass of cells, meaning glial cells, occupying much-needed space and consuming a high proportion of sugar and oxygen that communicating neurons could use for the benefit of our body.

Around 100 billion neurons are busy working for you 24-7 while around 850 billion glial cells seem to be taking a long nap. However, some glial cells out of that huge number of cells are part of our brain-blood barrier. They are performing as inspectors of blood being delivered to all the cells in the brain. There are other equally important jobs glial cells are responsible for. However, as far as I am concerned, it is a comparably small number of neurons that keep me busy learning about myself.

Both consciousness and unconsciousness weigh heavily in my thoughts. In my book *An Episodic Toxic Memory*, a survival story, I attempted to explore both subjects, but unfortunately, I came up short of a satisfactory answer. I hope neuroscientists can tackle the issue and provide a satisfactory explanation for most of us. You may suggest that research should begin with single-celled organisms, which it has. Peter, sitting at the end of this working table, may rightfully argue that single-celled organisms do not have the complexity of a human brain. John, a faithful believer in a supreme being, might intervene in this discussion by arguing that science can never decipher the mystery of the brain, as it is the home for man's soul, which cannot be placed on a petri dish or test tube, or be seen through a PET or fMRI scanner. This is an argument I prefer not to get involved in. I have attended several meetings on the above subjects. I have agreed to attend more meetings, but in the ones I have attended thus far, the opinions of the attendees at the end remained as far apart as when we began on the first day.

Jose Morales Dorta, PhD

You Are Unique in Our Solar System

I have no doubt that humans are the most beautiful and intelligent creatures inhabiting our solar system. It took many years for the cells making your body to develop and build the most complicated and beautiful organism existing in our solar system.

We are an intelligent species with vast knowledge about ourselves and everything existing on planet Earth. We know a lot about our sister planets and our sustaining star, the sun. Astrophysicists have calculated that we have a few more billion years to live before the sun changes states. We have developed sophisticated tools to examine and study ourselves, including the most complex organ in our galaxy, the human brain. This is a little far-fetched, but at present no one has challenged it. Just a decade ago, we completed the sequencing of the human genome, something that not even gods have attempted to do. At present we are engaged in a scientific research race to find deficient genes with variants and DNA markers that are linked to diseases.

In the past, humans pleaded to gods, shamans, healers of all kinds, and even religious icons for help. Today we have invented first-rate laboratories and sophisticated machines to probe the atoms, molecules, and cells of our bodies. We have made progress, but much remains to be learned about the intricacy of cell life in our organism. Naive promises have been made to the public—especially to individuals and families with chronic degenerative diseases hoping for a miracle pill. They were hoping to end their miserable and painful lives with new forms of scientifically proven drugs or therapeutic tools. Locating the gene or genes responsible for the etiology of our diseases has been quite elusive. However, we have made significant progress in all areas of medical technology and science in general in the last fifty to sixty years.

Committed and Enthusiastic, but Naive

Rather naively, but fully committed and enthusiastic, people hoped that diseases like Parkinson's disease, Alzheimer's disease, multiple sclerosis, and, to a lesser extent, even schizophrenia, bipolar disorder, and cancer, would be eradicated from planet Earth. During the 1990s, many predictions were made by laypeople as well as some scientists. However, it proved to be poor judgment and unproven science. Hardly any substantive treatment progress has been made or found on the above-mentioned human maladies.

Further hampering our scientific race, funding was cut in many areas of research. Moreover, concerned citizens negatively influenced congressional legislators over monies for stem cell research. Premature expectations and limited funding notwithstanding, the scientific community working in the sequencing of the human genome did not expect to find that most of our genome is composed of noncoding genes. The problems that presented to them continue to be unsolved. Even after discovering the so-called junk DNA, researchers did not know what to make of it. A few theories have been advanced, but the question remains largely unanswered. Similarly, DNA markers, epigenetics, and much more have slowed down our race in the hunt for the origin of most human diseases. There is a mountain of specialized research that needs to be translated into therapeutic tools and drugs. I applauded Francis Collins from the NIH for taking wise steps in this direction. Mel Graves, a cell biologist, wrote in *Nature*, "We fooled ourselves into thinking the genome was going to be transparent, a blue print, but it is not. Instead, sequencing and other new technology spew forth data, the complexity of biology has seemed to grow by order of magnitude." [63]

In the same article, Eric C. Hayden wrote, "Much of the noncoding DNA has a regulatory role, small RNA of different

---

[63] *Nature* 464, no. 7289 (April 1, 2010): 664.

varieties seem to control gene expression at the level of both DNA and RNA transcripts in ways that are still only beginning to become clear." (Please take note of the words *seem* and *beginning*.). The author of the article, E. C. Hayden, quoted another scientist, a mathematical biologist, who said, "Just the sheer existence of these exotic regulators suggest that our understanding about the most basic things- such as how a cell turns on and off- is incredibly naive."

In contrast, some of our cartoons make it easy for us to continue to entertain unrealistic promises. One song I heard in a cartoon goes like this:

Triplets nucleotides UGG codes for tryptophan;

it loves its neighbor AUG methionine with three letters each.

Not far away we find leucine that can come out in six different triplets.

All these letters are translated into amino acids that will become proteins for my body.

I found the song very interesting and educational. But I could not help stop thinking about how far removed from the reality of the process it is. Protein synthesis does not come as easy as presented in the central dogma of the double helix discoverers, Watson and Crick.

Protein Encoding Problems

The song mentioned in the previous section goes on to name more nucleotides and amino acids. It deserves our admiration and recognition for introducing the code of life, and thus biology and genetics, at an early phase of life. I wish I'd had a science-minded teacher during my early school years. However, the reality is not as simple as the song would have one believe. The genetic instructions of our complex organism have proven to be more difficult to decipher than we had anticipated. Nucleotides encoding for our proteins encounter myriad problems still largely unknown. Take, for instance, introns—the noncoding portion of our genome. These introns have to be identified, cut, and removed from the mRNA precursor before they can be considered for protein synthesis. Neither the noncoding genes nor the splicing process is understood well enough to be related to disease genesis. Furthermore, it is easy to talk about replication, transcription, and translation processes during protein manufacturing. However, connecting it to a syndrome or disease is basically an unrealized dream on the horizon of medical research. Nonetheless, DNA technology has become a strong and helpful ally of medicine and health in general.

For us, removing introns from exons in pre-mRNA is seen as a wonderful trick of our cells. However, further research in the protein manufacturing process is necessary for the benefit of disease genesis clarification. Considering pre-mRNA's first step and all the steps in between until we get mRNA and tRNA transported to a ribosome, culminating in very specialized proteins in the prefrontal cortex, it is a very long trip. It has to be celebrated as a wonderful achievement by our primordial cell.

The same thing can be said about specialized proteins in major organs of our body. Some researchers have voiced their concerns with identifying a locus or a simple step during protein synthesis that would end up in the production of a deficient organ or tissue. This concern stands out without considering many other causal factors in disease genesis. For instance, our laboratory companion, the mouse, is extremely helpful and relatively easily engineered to carry out our experiments. However, a schizophrenic or bipolar mouse, despite its almost similar genome, is not a human being that grows up with early symptoms that later on will develop into full-blown schizophrenia. The intricacy of a gene generating a multitude of messages has researchers wondering about a simple presentation of the human genome and undelivered past promises. They are working very hard, but the task ahead is a challenge in need of bright brains and commitment.

I have been focusing on rare variations that apply to populations worldwide. How about relatively new mutations, modifications, and adjustments that boost human evolution? Those mutations and changes might be in the process of making proper and adequate accommodations in our brains. Our hippocampi and prefrontal cortex are composed of groups of neurons under constant, persistent, and damaging stressful conditions. Residents of urban areas are highly exposed to stressors not imagined by our own great-grandparents. The entire human body, but in particular the brain, is making relevant adjustments to multiple demands to be able to survive. How exons and introns interact in critical regions of the brain to come out with appropriate signals for a timely and efficient response is another question that keeps on bugging me. Introns and glial cells are demanding more attention from all researchers. In addition, these researchers are not only referring to a specific locus in a cell or organ; they are contemplating the influence of an incomplete or wrong signal as it relates to healthy and not-so-healthy genes. They have in mind syndromes like schizophrenia, autism, and bipolar disorder. They see these not as a cut-and-dried diseases but as mountains of symptoms contributing to very painful brain disorders.

David Lewis and Schizophrenia

Leading researcher David Lewis has spent the past two decades exploring schizophrenia's developmental roots. He runs one of the most comprehensive and sustained attempts to explore normal and preschizophrenic adolescent brains. Dr. Mayberg, at Emory University, described it as "One of the smartest, most creative and most promising angles I know of on schizophrenia." Dr. Lewis focused his attention on the prefrontal cortex. In above paragraphs and chapters, I have shown you how this region of the brain is connected. It receives myriad projections from the limbic system, including the amygdalae and hippocampi, as well as subcortical and neocortical input. It is often called the CEO of the brain, which seems partially true because it is dependent on emotions coming from the limbic system to make wise decisions. Furthermore, the lifesaving amygdala has a mind of its own when it comes to protecting you from immediate danger.

Dr. Lewis's work examines the relationship between two types of cells: pyramidal neurons and chandelier cells. He thinks that chandelier cells fail to cultivate pyramidal cells during childhood or early adolescence. The communication needed in this region to develop a robust connection leaves the prefrontal cortex incapable of coordinated firing and synchronicity. Tom Insel from the NIH says Lewis's model provided something this field really needed—a framework for linking observation at the molecular, cellular, and system levels.[64] Coordination and cooperation among researchers is paramount in tackling syndrome-like diseases. Regarding brain diseases and disorders, I hope researchers do not fail to include the functions of the majority of its cells—the glial cells. Astrocytes seem to be eavesdropping on neurons at all time. Oligondendrocytes are responsible for producing myelin on neuronal axons, among other roles. There are many brain diseases connected to myelin deficiency.

---

[64] *Nature*, Vol. 468 (November 11, 2010): 155–6.

Atoms and Molecules in Action

WHEN WE OBSERVE THE multiple organelles within a cell, we wonder how atoms and molecules interact with each other to function normally. Molecules are constantly bouncing against each other in an aqueous environment in animal cells. The same must hold true for atoms, but it is very hard to advance an intelligent opinion on this elusive and microscopic particle of every element. Atoms and subatomic particles are constantly traveling at extremely high speeds through, above, and under me.

You are not exempt or free from this constant bombardment of atoms and subatomic particles either. Collisions among molecules and atoms of all sorts within and outside our cells produce force, meaning energy. My friend Pete was wondering if the abundance of force floating around molecules of our basic elements, such as hydrogen, carbon, oxygen, and nitrogen, makes those molecules exposed to degeneration. When they are degraded into their basic atomic components, they may provoke some of our diseases. This intelligent observation is an interesting one and deserves serious consideration by concerned researchers. Of course, there are multiple chemical reactions going on inside our cells that are necessary for normal function. Our bodies are in a continuous process of self-renewal (apoptosis and cell division and multiplication), but here I am referring to processes beyond the normal aging process. Or is it naive for me to separate these processes from one another?

Regarding self-renewal, I am thinking in line with our basic molecules during the formation of covalent bonds. These bonds are many times stronger than floating molecules or atoms. Our macromolecule DNA, with the assistance of another equally powerful molecule, RNA, is of paramount importance to you and me. Both molecules are engaged in the production of proteins called enzymes. These enzymes are responsible for the making and breaking down of covalent bonds. In other words, these enzymes can tear down the union of indispensable units of our organism that are built to tolerate a lot of tension and stress under critical conditions. They can come under attack by an enzyme from an incomplete or dysfunctional gene. This is a very valuable truth to remember while keeping an eye on your health. There are many genes in the human genome called introns; at present we do not know their function or functions. There are more noncoding genes than protein-producing genes. You already know that introns are cleaned out before messenger RNA is ready to be shipped out to become part of a protein. However, you also know the multiple problems and failures a protein encounters before its final step is completed.

Covalent Bonds

Please allow me to back up a bit to make my point clear to you. I'll begin by saying that covalent bonds are made when atoms share electrons. You may recall your elementary school teacher telling you that atoms are composed of protons, which have a positive charge, and neutrons, which are electrically neutral. Protons and neutrons and are located in the nucleus of an atom. The teacher likely also taught you and that atoms have electrons traveling around the nucleus in an orbital fashion. Electrons carry a negative electrical charge. Hydrogen has a positively charged proton with an electron, but it has no neutron. The most abundant atom in the human body is the hydrogen atom. It is the lightest element that is part of the body. Hydrogen atoms combine easily with other atoms. A basic law of physics to remember is that the electrical charge possessed by each proton is equal to but opposite by that carried or possessed by a single electron. The positive and negative charges of protons and electrons make the atom electrically neutral. The mass of each neutron in the nucleus is equal to that of each proton. This is another basic rule you should remember.

Another abundant atom in our body is carbon. In the nucleus, a carbon atom carries six protons, which are electrically positive charged. In addition, there are six neutrons. Most of the carbon atoms on planet Earth are composed of six protons and six neutrons and are known as carbon-12. Neutrons contribute to the stability of the atom, but the subatomic particle that definitely determines the element each atom belongs to is the proton. There is carbon-13 and carbon-14, both of which are known as isotopes. Carbon-13 has the normal six protons but seven neutrons. Carbon-14, a very radioactive isotope, has the normal six protons, but eight neutrons. The six protons in carbon-12 in its normal condition remain unaltered. If for any reason carbon-12 degenerates to have only five protons, or if it acquires an extra proton, making seven, that atom is no longer a carbon atom.

A couple of pages back, I mentioned covalent bonds. These bonds are very strong and do not break down easily. There is a significant difference between sharing an electron between two atoms that fulfill mutual specific needs for each other, and releasing or donating electrons to atoms or molecules. Besides strong covalent bonds, there are hydrogen bonds and ionic bonds, which I will not address here. While Watson and Crick were working on the structure of the double helix, one of the issues they faced was how the bases inside both backbones came together. Nucleotides from both backbones were united by hydrogen bonds, which are a weak type of bond. This hydrogen bond would facilitate the opening of the double helix for DNA replication and transcription. I brought this up in an effort to make you aware of the many complex processes our proteins, organs, and tissues go through before they become functionally normal in the body.

Nowadays, some individuals refer to us as being made of carbon. What this means is that if we were to do the impossible—meaning take all the water out of our body—nearly all the molecules in every cell in my body and your body would be based on the carbon atom. Carbon atoms are extremely versatile. They can easily form covalent bonds and join other carbon atoms to form large chains and rings for complex molecules. In the DNA double helix, the inside bases adenine, thymine, guanine, and cytosine are in the form of rings. Carbon atoms are involved in DNA methylation and histone modifications, among many other interventions in protein synthesis and normal cell functions. In the same vein, amino acids are a particular class of molecules that are distinguished from any other molecule in the human body. They all have an amino acid group (NH_2) and a carboxyl acid group linked to a single carbon that is commonly called the alpha carbon. There is a side chain or molecule known as R, which is attached to this alpha carbon. This side chain provides chemical variety to amino acids. I will not shy away from repeating that proteins are manufactured from twenty amino acids that are vital to life on Earth. It is not intended to be a joke, but amino acids and proteins are parts of your body in sickness and in health.

Just as amino acids are the subunits of proteins, nucleotides are the subunits of DNA and RNA molecules. Basically, a nucleotide is a molecule compound of a nitrogen-containing ring compound linked to a five-carbon sugar. The sugar in this nucleotide can be either ribose or deoxyribose; it carries one or more phosphate group. You see it clearly in the double helix as per Watson and Crick's classical illustration. Although the word *nucleotide* is commonly used to define or refer to the famous letters A, T, C, and G, it also refers to the sugar-phosphate backbone segment linked with each base pair on both sides of the double helix. Nucleotides have additional functions I am not equipped to deal with here. In summary, there is the ribose-containing molecule RNA and the deoxyribose molecule. The RNA molecule spews out uracil for an exchange of thymine, and DNA equally spews out A, T, C, and G.

Consider the repetition of this material an innocent attempt on my part for you to remember me. My teacher used to hand out handwritten declarative sentences like "On a weight basis, the macromolecules DNA and RNA are by far the most abundant of the carbon-containing molecules in a living cell." To make her point even more clearly to us, she used to add that all of the cells in her body were made of carbon atoms. Carbon molecules are the most distinctive properties of all living things. At the time, I was wondering about my soul, oxygen, nitrogen, hydrogen and other elements that existed in my body. However, she did not emphasize these as much as she did carbon atoms. I remember her as a caring and dedicated teacher who accomplished her goal of creating long-term memory in my brain. She hammered on DNA and RNA molecules because she knew, among other things, that both molecules are engaged in the formation and storage of memories, in addition to the transmission of hereditary information. And involved in this storage and transmission of information is the risk of inherited disease I have been writing about.

It is truly a long journey from our ancestors thousands of years ago to the present. We have survived not only the jungle but also armies of microbes, plagues, and natural disasters of all kinds. In great urban areas, when we ride public buses and subways, we are breathing the same air as sick people riding with us. A secondary problem coming out of sharing the same air trapped in subways and elevators is that not everybody has developed the same defenses against bodily intruders like viruses and bacteria that will make us sick. I was born and raised out in the country, not far from the ocean. My family was used to eating fresh fruit and vegetables grown on our small farm. We ate many beans and roots, accompanied by homegrown chickens, goats, and cows. We usually dried and salted the meat at home. My mother, like her father, was a locally well-known botanist who kept a medicinal garden at home for the family's and neighbors' needs. Although tobacco was one of the main sources of income in our home, none of us learned to smoke cigarettes; it was exported up north. I do not remember going to a medical doctor, except a dentist, until I was nineteen years old. When I left my country home and moved to a northern urban area and began riding trolleys and subways, I was exposed to alien microorganisms my body had not built adequate defenses for.

Notes from My Teacher

I carry notes reminding me that the DNA polymerase is trusted with the responsibility of manufacturing new DNA strands by using the old strand as a template. This self-replication process provides the mechanism for the addition of new double strands of DNA molecules. DNA replication produces two double helices from an original DNA molecule.

Each of the two strands of DNA is used as a template for building complementary strands. On a blackboard, my teacher once wrote, "Most genes are short segments of DNA encoding a single protein. The DNA in a gene is not used in its totality in encoding a single protein. Short stretches of DNA become busy identifying and determining timing, place, and size, among other jobs, in protein manufacture. I do not tire of repeating that you have a very wise DNA molecule. Your genome is a large and effective information storage vault that deserves much respect and awesome admiration. Another awesome attribute of the DNA double strand and self-replication is that when the DNA separates for replication, the information is passed on to the complementary strand.

The DNA Replication Process

The DNA replication process is initiated by an enzyme that attaches itself to the DNA and unzips and separates each strand. It does this job by or unbinding the hydrogen bonds between the bases. There are particular gene segments responsible for signaling where the replication process begins. During the whole process of replication, there is a very wise enzyme, known to us as DNA polymerase, that is busy directing this complex operation. It synthesizes the newly formed DNA strands using one of the old strands as a template. This wise DNA polymerase commands the addition of nucleotides to the growing DNA strand. The term "old" does not refer to time or age; it refers to one of the original strands that will serve as a mold for a complementary new strand. For this very reason, each strand of a DNA molecule becomes a template for a complementary strand. This enables the cell to copy its genes before passing them on to coming generations.

This wonderful process that is carried out within the cell must be done extremely quickly and accurately. The beauty of this replication process is that it may include many thousands of nucleotide pairs. This replication machine, which is basically composed of proteins, manufactures two complete double helices that are identical to the original DNA. This replication process, which involves thousands upon thousands of nucleotides, must occur with a minimum of errors. Take a minute to think about people who live to be a hundred years old. In them, those molecules responsible for replication, transcription, and translation are active from birth to death without taking a week or two of vacation per year; they deserve to be respected and praised for a job well done.

Your Chemistry: DNA

By getting to know your chemistry, meaning your DNA, you will learn the rules that govern all forms of life on planet Earth. A chemical soup gave rise to an RNA and DNA world that—in relative harmony with a wise organelle, the mitochondria—has produced the most intelligent organism in our solar system. Our laws of physics, chemistry, and biology seem to work in unison with, and are applicable to, the rest of the universe. No matter which element of the human body you look at, it is composed of atoms. And atoms of various elements are traveling through space in the form of clouds and gases, and they become attracted to each other and form celestial bodies like stars. And stars, like humans, are in a constant dynamic process of birth and death; this has been so ever since the big bang took place somewhere around 18 billion years ago. Some stars go through the process of implosion-explosion, becoming what is known as a supernova. This is a recycling process in which the old star goes through a cycle of self-consumption to appear anew in atoms and molecules in another space and time. A few billion years ago, our star, the sun, was born; and you can rest assured that in another few billion years, it will die. You and I will not be here to witness it. However, the same atoms—the same energy that exists in both of us at the present time—will forever exist (unless it falls prey to a black hole, according to most modern astrophysicists).

Tiny Particles inside the Atom

In an atom, there are tiny particles and waves responsible for chemical and atomic processes we are beginning to understand and use for our own benefit. Likewise, we have chromosomes in our bodily cells that are structures within the cell nucleus where genes are carried or stored. There are more genes than chromosomes. It is not only a scientific truth but a logical conclusion. Consider the genes you must carry for the colors of your eyes, skin, and hair. Even your height, to some extent, is determined by your genes. These genes you get from your parents are filed in your DNA, the great and wise molecule, where they await expression during mating.

The genes you have inherited from your ancestors do not fade away, fuse, or blend in a fertilized egg. They remain unaltered, or as Gregor Mendel called it, particulate, meaning that those very particular genes are not lost after fertilization. This wise and humble dedicated monk, G. Mendel, discovered the unbending quality of genes around 150 years ago during his spare time while teaching at a monastery. I can say without an ounce of doubt that he did not have a first-rate computer to work out his mathematics and necessary computations for his experiments. Monasteries during the second half of the nineteenth century in Eastern Europe were not known for the abundance of anything. However, he must have managed to keep paper and a pencil or pen in his dormitory room.

The monk and geneticist Mendel knew that his discoveries were very important, but his fellows did not appreciate it. In 1865 he gave a lecture on his work at a local medical society, followed by publication of his discoveries in the society's journal. Unbelievably, his findings remained unknown to the scientific world and public in general until the first decade of the twentieth century. Mendel had proven scientifically what farmers had been practicing for thousands of years. Farmers had been selectively using the best and most desirable offspring of animals, birds, and plants to improve their stock. It appears to me that Mendel's work has just recently begun to be fully appreciated, during and after the sequencing of the human genome.

We are learning to engineer genes for medical and food research for the benefit of humanity in general. A large section of this essay has been centered on DNA, genes, and proteins. Gregor Mendel was born in 1822. He did not live to see the fruit of his work appreciated beyond the walls of his monastery. For me and many others who love his work, Mendel is another hero who deserves to be paraded along the Canyon of Heroes on Broadway in Manhattan. Without insulting anyone, the science of genetics was born by the counting of round and wrinkled peas by a monk who was not known by his intellectual acumen.

I am considering including G. Galileo, Marie Curie, and Rosalind Franklin on the list of outstanding scientists to be selected for the Canyon of Heroes. You should have a voice in the selection of our heroes. I did not mention Francois Jacob and Jacques Monod, because they were paraded along the Arc de Triomphe in Paris.

Gregor Mendel reached his conclusion by observing hereditary determinants—genes—as phenotypes, meaning the outside appearance or expression they resulted in. With pea plants, he observed and counted spherical and wrinkled peas. He began his work by selecting and observing purely spherical and purely wrinkled peas. He paired or fertilized a few of these so-called pure breeds. The first generation of fertilized peas, which are classified by geneticists as F1, came out all spherical. He then paired or fertilized the offspring of the first and subsequent generations of his selected peas. He found out that they came out as spherical and wrinkled peas in a ratio of 3 to 1. He called the spherical traits in the peas dominant. He called the wrinkled peas recessive hereditary determinants. What Mendel called hereditary determinants are known to us as genes.

Without tools to look inside each pea to examine and analyze how hereditary determinants work and remain "particulate," Mendel had half opened the door into reading the book of life. Future scientists became interested in discovering how and where hereditary determinants—genes—are formed and stored. Scientists wanted the key that would enable them to understand how peas, flowers, plants, animals, and, above all, human beings are built from, apparently, two simple molecules. I never tire of thanking this simple and humble monk for showing us, among other things, the outside expression of genes.

After this, researchers became engaged in an investigational race for all sorts of genes. These include mutated genes, disease-provoking genes, rare gene variants, and many others. They wanted to identify, characterize, and look inside genes and use their millions-of-years-old code for the benefit of humans. After a short hesitation to determine whether hereditary information was stored in proteins, amino acids, or similar chemical structures within the cell, scientific experiments proved without any doubt that the genetic code resides in the DNA molecule.

Among the big players during this hunt for genes were: L. Pauling, known to many scientists as the world's leading chemist during his lifetime. Besides Pauling, there was Rosalind Franklin, Maurice Wilkins, and, of course, my intellectual icons J. D. Watson and Francis Crick. I am not overlooking the contribution made by the world's best-known neurologist, Santiago R. Cajal and so many others it would be impossible for me to list them all here. Cajal made it clear for us that the human brain is not a long sausage coiled around itself. He made clear that our brain is composed of neurons and glial cells that are independent of each other. When Watson and Crick first introduced the double helix, they stated, "It has not escaped our notice that the specific pairing we have postulated ..." Their statement changed biology from the cataloging and classification of animals, plants, and birds into a scientific race rivaled only by the race to understand the atom. From that date on, presidents and prime ministers on both sides of the Atlantic began to appear on the front pages of newspapers and scientific journals with Nobel laureates and pioneers in biology.

Watson and Crick had set in motion a race among scientists around the world. They wanted to know how amino acids were manufactured and built into proteins. The genetic information Mendel was referring to was found to be encoded in the sequence of nucleotides—the letters of DNA. However, a clear and accurate look at DNA and how it determines its nucleotide formation still needed some extra work and delicate investigation. At the core of this scientific investigation and intellectual curiosity lies the way genes work and behave at molecular level. Genes, proteins, diseases, drugs to cure those diseases, and research tools seem to have occupied the minds of scientists around the world. Out on the horizon, far away and hardly conceptualized by most people, there were scientists attempting to understand gene behavior and the sequencing of the human genome. Questions such as How do we look at uncoiled or unwrapped DNA strands and the chromosome itself, were academic debates, with very few scientists engaged in laboratory work. Researchers wanted to know how long strands of nucleotides coiled up in bead-like structures with multiple amino acids could conserve Mendel's particulate quality.

DNA Technology and Restriction Enzymes

THIS RACE IN BIOLOGY was turned into a technological revolution relating to DNA and RNA. Biologists, chemists, physicists, mathematicians, and computer engineers began to see the advantages of working together. Among the first revolutionary technological DNA tools were restriction enzymes. The discovery of these enzymes allowed scientists to cut DNA molecule segments. DNA segments cut using restriction enzymes—meaning single genes or clusters of genes scientists were interested in—were isolated for study, analysis, and purposeful scientific use. There was a scientific goal in mind to achieve using this new technology. The next step scientists took upon themselves following the discovery of restriction enzymes was the use of an enzyme dubbed *ligase*. The ligase enzyme could be used to paste or glue together DNA sequences—genes—for whatever reason the researcher had in mind.

According to J. D. Watson, there was a young scientist gifted with a mind for business and an enthusiasm for the genome sequence. "Herb Boyer, was already an expert on restriction enzymes in an era when hardly anyone had heard of them," wrote Watson. At this early stage of DNA technology, we had discovered two enzymes to do work for us. What wonderful discoveries! Mendel had played with phenotype expression of genes, but now scientists were touching and manipulating the genes themselves. Furthermore, the technology of cutting and gluing genes together provided scientists the ability to create plasmids. A plasmid is roughly acts like a CD, recording and storing information. By luck or coincidence, another brilliant scientist, Stanley Cohen, attended a conference on plasmids in Honolulu in 1972. So, by the grace of curiosity and scientific interest, S. Cohen and Herb Boyer, the restriction enzyme star, realized that together they had a DNA technology nobody else knew about.

Aside from their interest in genes, the two seem to have worked very well together, and they established the first recombinant DNA company in the world. This was the world's first genetic engineering establishment. It was founded on the US West Coast. This game that consisted of playing around with genes and enzymes was done with small animals. The big job, the sequencing of the human genome, was resting on the drawing boards of optimistic biologists and scientists.

Boyer and Cohen's discoveries and DNA technology tools were not small things that went unnoticed by the scientific community around the world. Scientists could now engineer bacteria and small animal cells, making cuts with restriction enzymes and putting gene segments back together with a ligase enzyme. Cutting and joining together compatible segments of genes gave birth to another novel idea—recombinant DNA technology. One dear parasite in our intestines, E. coli, proved to be an excellent subject to work with. It was cheap, and it demanded very little. Determining how to transfer DNA material from one organism to another was the next challenge, so scientists used plasmids to do the job. The question that lay ahead was, how would the host receive its guest?

When foreign dignitaries come to Washington, a big red carpet is displayed for them. Not all of them deserve such a privileged treatment, but politics has its own rules of behavior. You already know how an organism tends to defend itself and reject what it considers invaders or intruders. I am referring here to the immune system. There is a group of defensive cells that has its origin in the bone marrow, with T cells maturing in the thymus. It is a survival system that has defended us for thousands of years. It mobilizes thousands upon thousands of T and B cells to keep people alive when toxic invaders try to break into their bodies. What I am saying is that the solution to recombinant DNA rejection already exists in the human organism. Your brain has its own blood-brain barrier to filter out undesirable organisms.

Boyer and Cohen had the solution at hand—using a plasmid. Cuts on sequence segments made by plasmids would be glued together with a ligase enzyme. At first glance, this did not provide much help to Boyer and his plasmid pioneer. The ligase enzyme was doing what was already known to the researchers—gluing stretches of plasmid together. However, the opportunity to come out of the laboratory with his test tube in hand, crying out, "Eureka!" was not far away. During a following test, it was observed that the gluing enzyme would manipulate a cut plasmid to incorporate segments of DNA from a neighboring plasmid, thus creating a hybrid. It was in Cohen's hands to use his proven technology to transplant the whole plasmid into a bacterium for multiplication purposes. Resistance to the newly added guest was overcome using already available antigen resistance. The next step would be the creation of a hybrid from two different species.

An abundance of flies and birds can be captured and used in laboratories for the benefit of mankind. I remember crows coming to my father's cornfield to eat our corn despite our strong objections to it. My father spent money and time planting and cultivating corn, while the crows had a feast at our expense. We could have captured a few crows, extracted a few cells, cut DNA sequences using restriction enzymes, glued those sequences to a bacteria plasmid using a DNA ligase enzyme, and bingo, we would have had a recombinant DNA molecule. It looks very simple the way I have described it here. Perhaps if we practiced it five or six times, it would be a simple job for all of us. It would be just like eating a piece of cake.

TV Shows and DNA Fingerprinting

Up to this point, most of the work I have been writing about was done with bacteria. The next step would be done with higher animals having a genome similar to ours, like our little brother the mouse. Recombinant DNA technology is and will be an excellent tool for the discovery of DNA mutations that are responsible for inherited diseases. It is already a first-rate tool in medical research and therapeutic tools beginning with insulin for diabetics (Boyer and Cohen) and blood-clotting proteins. Another tool I cannot ignore in this essay is DNA fingerprinting, although the finger is hardly involved in the game. Sometimes I enjoy television shows presenting two or more young fellows claiming or disclaiming paternity privileges or monetary contributions to the mother of a child. At times I believe the TV show host presenting two or more males making the same claim is just a charade to arouse the spectator's feelings and improve the show's ratings. However, DNA technology has come out to be the jury and judge in cases of declaring who is the father of a child in question. DNA identity is 99.99 percent accurate, despite a mother's allegations that she slept with one of the fellows only two times. In addition shows that present paternity issues, I like to watch shows that present crime stories in which DNA technology is involved in solving a crime.

Jose Morales Dorta, PhD

Attempts to Hide DNA Fingerprints

There have been criminal cases that were extremely difficult for police officers and detectives to solve. In such cases, the perpetrator is normally a cold and calculating person who has carefully considered all possible clues he might have left behind at the scene of the crime. Police officers and detectives might refer to him as a professional criminal whose specialization is eliminating all traces that could possibly lead to identification.

Among the things such a criminal might do are covering footprints in multiple ways and changing his appearance by wearing a mustache. Sometimes the criminal agent paints or enlarges his eyebrows, modifies the shape of his nose, or alters his voice. He may appear to be the most elusive serial killer, rapist, or bank robber. Such a criminal might be placed on a list of most wanted criminals and may have hundreds of agents watching for him based on a specific detailed description. He might have changed his phenotype, or outside appearance, but his DNA profile is forever; no changes are allowed. This criminal might have forgotten that he left his DNA on a cup of coffee, soda can, or cigarette butt, and that could be enough to put him behind bars after he has seen the awaiting judge.

Leaving behind the criminals and paternity claimants, let us return to our wonder-working enzymes. Restriction enzymes cut DNA molecules at specific sites. They know exactly where to go, assisted by small segments of genes. Those cells of yours are not stupid at all; they have developed many tricks for self-protection through the years. We are just now learning about these things. While our scientists were doing their jobs investigating the inside of bacteria, they observed that the host bacteria cut plasmid DNA segments at particular places. The mysterious enzymes (restriction enzymes) cut DNA only at certain nucleotide sequences. This was cause for another worldwide celebration.

The restriction enzymes were very particular in choosing the sites at which to do the cutting. Moreover, there was another very interesting factor that had to be examined and included. Not all bacteria possess the same restriction enzyme; therefore, some restriction enzymes cut at different sequences of nucleotides. This discovery was extremely important for the near future. It played a critical role during the sequencing of large genomes. One of the rules of the cutting enzymes is that they choose to cut short segments, from four to eight nucleotide pairs. It followed that because there are different bacteria, each species or subspecies would produce unique enzymes that cut at specific sequences. These discoveries regarding enzyme cutting proved very useful for the future sequencing of large genomes.

Restriction enzymes cut at certain sites containing different amounts of base pairs. In a properly and adequately prepared aqueous medium or gel, the longest and heaviest cuts of base pairs would occur at places different from the shorter segments of base pairs. It sounds logical and even practical to me that scientists came to that conclusion at the time. Brain power plus dedication and perseverance make scientists a breed apart.

The abundance of bacteria in the environment provided multiple restriction enzymes that could cut at various sites and were ready for use in laboratories. Engineers in this field of molecular biology take specifically tagged nucleases from among the multiple restriction nucleases available.

Electrophore Technology

Restriction enzymes will cut a given DNA molecule at the specified site intended by the technologist. DNA fragment separation is accomplished, in most cases, using gel electrophore. It will separate fragments or base pairs within larger fragments. One relatively easy and new way of doing this is carried out by staining DNA with a dye that fluoresces under ultraviolet light. (On the ocean beach in Lajas, Puerto Rico, when the sun gives way to the night and the stars adorn the sky, you can see thousands upon thousands of tiny fish carrying a fluorescent protein that sets them apart from any other fish. Take a short vacation and visit it; you may be motivated to do some type of research with those beautiful and interesting fish. If a tiny fish can develop that kind of technology, your brain can do even better.)

Another model used by scientists is to incorporate radioisotopes into DNA molecules. Cold Spring Harbor Laboratory on Long Island in New York is a pioneering center of research in this field. My dear friend J. D. Watson made his first presentation of the double helix there.

Recombinant DNA

DNA technology uses the most advanced equipment available to commercialize its products in many journals. With the tools available to DNA technologists, long DNA molecules can be cut to the desired sizes of fragments for an intended job. You just need to then join the fragments together with the enzyme I have already mentioned to you. You do not need to worry about the dogma developed by Watson and Crick that strands will attract each other to form a double helix based on chemical attraction. The nucleotide adenine will attract thymine, and cytozine will attract guanine. You do not need to worry about this while cutting and joining DNA fragments.

The DNA chemical rules governing all organisms permits DNA from practically any animal source to be joined together. This chemical process allows technologists to have free reign in cutting and joining DNA fragments from different organisms. The enzyme ligase can join together two fragments without asking questions about their origin or place of birth. And the energy to do the job is available and plentiful in the form of another protein, the ATP molecule. It will bring together the sugar-phosphate backbone. The backbone fragments are identified with the digits 3 and 5 in most illustrations. The 3 end of a backbone has the shape of a half moon or the letter C, while the 5 end is spherical. When joined together, they will fit perfectly into each other. You do not need to use a pair of pliers to do it; enzymes will do it for you.

During this apparently simple process you will be playing with chemical attraction of base pairs. The energy-providing ATP molecule, restriction enzymes, and DNA ligase enzymes will also participate in this game. When these processes are taking place inside your body, you will have supervisors and guardian proteins watching over the job in case there is an error. Your cells are always on high-alert status, ready to intervene when necessary. All these mechanisms fall within the code of life your cells have retained for your own safety. Please do not destroy it; it took your primordial cell a long time to build it.

The Laboratory Workhorse—a Bacterium

Base pairs in DNA segments are the same in all living organisms. Once you join the segments together as in a plasmid, the fragment will continue to replicate as if it were coming from a unique mother. Basically, bacteria seem to be our best workhorses for duplicating our DNA fragments. The bacterium is an excellent tool to copy DNA fragments several times quickly and in an economically prudent fashion. Every time a bacterium replicates or duplicates itself, it will also replicate the DNA fragment you inserted into it.

You may want to or need to replicate human DNA fragments, making use of another organism for future use. You may need to cultivate DNA fragments from a person to repair an ear or nose. I would like to share with you that this is not science fiction. When working with DNA fragments introduced into a bacterium, the fragments need some protection from hostile fragments; otherwise, they will become food for the bacterium. A DNA enthusiast at the time (Stanley Cohen) was nearby with a plasmid to be used as a carrier, also known as vector. This carrier model—a plasmid—possesses a replication original that will duplicate independently of the bacterium chromosome. During the manufacture of a final product, restriction enzymes and DNA ligase play major roles on the lasso shaped or circular plasmid in recombinant DNA. Recombinant DNA is copied as bacteria divide many times.

The final step during the manufacture of recombinant DNA is purification of the DNA from the content of the rest of the cell. This process includes the removal of the bacterium chromosome. The final purified plasmid DNA may contain thousands upon thousands copies of the original DNA fragment. There are sophisticated tools to get the DNA fragment out of the plasmid cleanly. The thing that impresses me most is that all this new technology was and is stored in your brain cells.

Just One More Time

One more time, I will continue to elaborate on DNA technology available for research at local individual laboratories. I stopped at restriction enzymes, plasmids, and recombinant DNA. I centered my attention on my pioneering icons Boyer and Cohen for many obvious reasons. There are other equally outstanding pioneers in the field I did not elaborate on in order to concentrate on DNA and its multiple tools. However, I feel obliged to write something on another Grand Canyon hero, Warner Arber, a Swiss biochemist whose main concern was the reason some viral DNA were degraded after being introduced into a bacterium. Those were days of grand pronouncements on the double helix, amino acids, triplets, and protein synthesis. They were subjects of investigation conducive to winning a Nobel Prize and having big celebrations in luxurious hotels. Arber's brain was busy at home wondering why some cells were prejudiced against some virus's DNA.

Host cells would selectively destroy some viral DNA. What characterized or distinguished that viral DNA from the rest of the DNA? A DNA molecule is the same whether one is talking about a microorganism like a virus, a bacterium, an elephant, or a sequoia tree in California. Arber's sleep was interrupted by his wondering whether an unknown molecular chemical process or something else was inside the cell that kept its own DNA safe but fiercely attacked viral DNA. He faced the challenge with great enthusiasm and an inquisitive mind. He discovered that the molecule in question was none other than the now very famous restriction enzyme. His discovery became a first-rate research tool in many DNA laboratories. It became the darling enzyme in the creation of recombinant DNA.

Restriction enzymes in bacteria cells restrict viral growth by cutting foreign DNA. This is the insightful moment that Werner Arber had been waiting for even when asleep. A lot of science work is done in one's head by proposing ideas and solutions to problems. Many proposed solutions end up in a laboratory or are rehearsed by peers and coworkers. Further investigation into restriction enzymes showed that DNA cutting is not a random job. It is a sequence-specific reaction. It does not cut capriciously. A given enzyme will cut DNA only when it recognizes a particular sequence.

Obviously, this was not a small and insignificant discovery that could have remained hidden among multiple notes and rusty tools. This type of discovery one that causes you to stop eating your sandwich, grab your cell phone, and contact your friend immediately.

Jose Morales Dorta, PhD

My Noxious Friend and Tenant E. Coli

I already mentioned EcoRI as being one of the first restriction enzymes to become a star among enzymes. It recognizes the six-letter sequence GAATTC and cuts it from the rest of the letters. Up to this point, it sounds and looks easy. But in practice, EcoRI saw a six-letters sequence of GAATTC, made its cut, and stopped cutting. How come EcoRI did not cut the genome every time the same sequence of letters appeared?

W. Arber was doing his science 24-7 and discovered that the bacterium he was working with manufactured another enzyme. This newly discovered enzyme did not allow automatic cutting of the six-letter sequence listed above. EcoRI's cutting privileges were restricted by another enzyme from the bacterium. This newly found enzyme modified the structure and chemistry of those six-letter sequences of its own DNA. This modification of structure and chemistry was achieved by adding a methyl group to the bases. The methyl group consists of one carbon atom and three hydrogen atoms. The cell itself had invented the cutting restriction enzyme, the sequence-cutting interval, and the chemical modification to protect its own DNA from being cut automatically. When EcoRI was tried to cut the prearranged sequences of letters, it did not recognize the six letters; otherwise it would have made its cuts without hesitation. It looks easy, and it should be. Bacteria have all these tools. We, as superior organisms, have them, and W. Arber proved this for us.

DNA-sequence cutting moved along quite well with new discoveries made on both sides of the Atlantic. Technological DNA jargon was increasingly heard in bars, restaurants, and even barber shops. For the next step, there was a problem that scientists and medical doctors were aware of—bacterial resistance to antibiotics. Some people were not responding to antibiotics because the medication prescribed did not kill the pathogens inside the body of the sick person. The invasive bacterium had gone through a mutation in its genome. The newly discovered antibiotic resistance was provoked by incorporating a small foreign fragment of DNA, later called a plasmid.

A plasmid has the shape of a lasso or loop. Plasmids have made bacteria their main permanent places of residence. This lasso/loop structure that lives within a bacterium is not a passive, inactive, and uninvolved piece of DNA. It possesses the ability to replicate and move on to the next step along with the rest of the bacterium's genome when the cell goes through its stage of division. The cell cannot leave it behind or get rid of it during cell division. This has been repeated several times; most cells have survived for millions upon millions of years by cell division and reproduction. First I explained that a bacterium, a cell, survived threats by mutation, and secondly I explained that cells have another survival tool—the plasmid. Surprises are not absent during investigating microorganisms like bacteria. Plasmids were found to pass from a single bacterium to another that originally did not possess it. It was a blessing for the receiving bacterium because it received DNA genes that provided antibiotic resistance.

It remained for Stanley Cohen at Stanford University to further elucidate for us the new antibiotic tool, the plasmid. As the story goes, Cohen and Boyer met at a conference in Honolulu in Hawaii. Both pioneers in restriction enzymes and plasmids gave birth to recombinant DNA. One of the first products of this union was insulin.

This stage of genetic engineering presented serious questions we had never encountered before. It meant we could play with genes at will, cutting, adding, and exchanging fragments of DNA among surprisingly different species. We manipulated genes from bacteria to bacteria, discovering and forming antibiotic tools for the benefit of many people around the world. We are composed of DNA.

There are frogs, mice, chickens, sheep, dogs, and cats we could begin transferring genes—plasmids—from for medical and research discoveries. The immense benefit these advances could bring to the agricultural industry—including improvements in vegetables, seeds, and roots—did not go unnoticed.

A second discovery of our planet Earth came about as scientists explored microorganisms never seen or heard of before. A new scientific language was able to address new frontiers in all areas of modern science. It opened doors for women to explore and succeed in this new field of research. We have traveled a long way from scientist Marie Curie to the present. Two of my neuroscientist icons, Nancy C. Andreasen and Jeanette Norden, are world-renowned professors whose teaching talents have motivated thousands of students on planet Earth. They are outstanding researchers and widely acclaimed university professors. I love to listen to their lectures and read their books. During all these DNA-technology discoveries, there was not a single Frankenstein working behind closed doors in a castle in faraway Ingolstadt. These were men and women exchanging information by phone calls, journals, and conferences all over the world. The molecular biologists, geneticists, chemists, paleontologists, physicists, farmers, and fishermen were turning their attention to this new technology. Viruses and bacteria were the objects of meticulous study for possible use in many areas of related investigation. Farmers the world over were interested not only in improving the output of crops but also in antibiotics for their seeds and plants. Our hero, J. D. Watson was already head of the Cold Spring Harbor Laboratory on Long Island, a hot spot for enthusiastic and highly intelligent young scientists racing to make their discoveries front-page news. Helping people is their main concern.

A Research Hiatus

E. Coli was a very hot tool of research not only in the United States and Europe but also in Asia. Dr. Watson was a cautious head scientist and was not in a rush to open Pandora's box. There was a five-year span of extremely cautious research, but the brilliant researchers closed ranks and went on doing first-class research.

Going from microorganisms to relatively large four-legged animals was no small jump. Inserting the desired small fragment of DNA into a plasmid to be transferred into a bacterium allowed the bacteria to divide, multiply, and grow, producing a large number of copies of DNA fragments. This step might be called the first step involving the sequencing of a large genome. And among large genomes, there was, in the minds of some visionary and enthusiastic scientists, the sequencing of the human genome. One more time, the Atlantic Ocean was the divider between two brilliant pioneers in sequence technology. Wally Gilbert was at Harvard University, and Fred Sanger was at Cambridge University in England. Sanger's approach to sequencing using the same enzyme that copies DNA naturally in cells, the now famous molecule DNA polymerase, proved to be the more practical of the methods the two were using.

Protein and enzyme functions

Protein and enzyme functions are manifold. Proteins and enzymes come in different shapes located in different organs and tissues of our body. Among their functions are responsibilities for building necessary tissue for repairing occasional bruises. Another function is fighting and destroying undesirable and toxic bodily intruders like viruses and bacteria. They have done a wonderful job throughout our evolutionary history of healing fractures and wounds sustained while fighting and capturing animals for food. Enzyme functions are worthy of admiration, praise, and careful scientific observation and investigation. There is an enzyme dubbed *Dicer* which, in another essay, I referred to as a horseman with a sharp sword. Among its functions, this horseman cuts long double-stranded RNA molecules into shorter pieces. These short pieces are crucial for gene-silencing pathways that involve small RNA, such as short interfering RNA or "microRNA, which are the most abundant classes of small RNA molecules."[65] This means the cutting enzyme Dicer is indispensable in generating the above-named small RNA tools. A deficit of Dicer enzymes is believed to be involved in macular degeneration, an eye disease related to old age. Previous work has shown that reduced Dicer levels can occur in many tissues and are associated with various diseases. Besides its crucial role in RNA-controlled gene silencing, mammalian Dicer1 (our bodies contain this enzyme) seems to have another function. "It maintains visual health by degrading toxic RNA molecules." [66]

Scientist Hiroki Kanek, one of the leading investigators at the University of Kentucky wrote, "Our findings elucidate a critical cell survival function for Dicer 1 by functional silencing of toxic Alu transcripts. (Alu RNAs are transcripts of Alu elements- the most abundant non- coding repetitive DNA sequences in the human genome.) This unexpected function suggests that RNAi-independent mechanism should be considered in interpreting the phenotype of system in which Dicer is deregulated." The author further alerts us by writing, "Recognition of Dicer1 unidentified function as an important controller of transcripts derived from the most abundant genome repetitive elements can illuminate new functions for RNases (enzymes) in cytoprotective surveillance."

The horse with a sharp sword I wrote about is not only involved in diseases but is actively engaged in the most abundant noncoding

---

[65] Bartel D. P., *Cell* 136 (2009): p. 215–33.
[66] *Nature* 471 (March 17, 2011): 308.

repetitive DNA sequences of our genome. You may recall that these are the same genome sequences we used to call junk DNA. Regarding these sequences, some congressmen complained that the federal government was wasting money on studying them.

H. Kanek further cautions us, writing, "To our knowledge, this is the first example of how Alu could cause a human disease via direct RNA cytotoxicity rather than by inducing chromosomal DNA re-arrangements or insertion mutation through retro-transposition which have been implicated in a thalassemia, colon cancer and neuro-fibromatosis." [67]

Following signals and functions of enzymes along pathways of multiple cell interventions has proven an arduous and challenging task for researchers, the efforts are paying off in many ways. A few years back, scientists were complaining about the excessive amount of nonessential glial cells occupying most of our skull. These glial cells were referred to as a left over from our chimpanzee heritage. Now we are learning they are involved in many unsuspected roles and necessary for normal brain function. The job of microRNAs is not fully established yet, despite its wide use as a research tool, but neither are the multiple forms and functions of Dicer. It is evident that they are involved in human diseases, but to identify the locus of a particular disease among billions of cells and trillions of synapses takes time and absolute dedication. Following RNAs during their multiple functions starting with protein assembly and including short interfering fragments cutting existing double strands is a huge job.

I have to add that the demands of the health care industry are difficult for researchers in that area to contend with. Knowing when to stitch polymers together for a specialized job and when to cut them, even with gene fragment help, is marvelous and ingenious by any standard. The job of manufacturing, degenerating, and

---

[67] *Nature* 471 (March 17, 2011): 325–9.

assembling units and repairing them at the micro level when the need arises is a big challenge in my book of notes.

Learning the Trade

Gene splicing led scientists to explore and master the technology needed to cut and splice nucleotide sequences, meaning recombinant DNA technology. Splicing or joining together any nucleotide sequence to build a new product for commercial purposes came about with the development of this new technology. Any genetic sequence can be cut using restriction enzymes and be spliced into another sequence of letters to manufacture a medical product for diseases that have remained undefeated until now. Restriction enzymes are used in laboratories to remove or to add nucleotides to sequences. This technology has facilitated pharmaceutical companies to manufacture large amounts of essential proteins for medical purposes. Most diabetics have benefited from this wonderful DNA medical technology. We can insert insulin-producing genes into the genomes of bacteria to produce large amounts of the proteins necessary for insulin genesis. Some individuals may refer to this as cloning insulin-producing genes.

I will make a final comment on this most significant scientific advancement that took place during my lifetime. It took place twenty-four hundred years after Socrates's death in Athens. During the processes discussed above, scientists discovered how their tool, a bacterium, had evolved to protect itself from a viral infection. The wise bacterium had learned to manufacture enzymes to recognize a specific DNA sequence on the invading virus and made a cut at that threatening sequence. There is an abundance of enzymes in nature that we need to identify and catalog. They can cut the specific sequences of toxic invaders. You just need the curiosity and enthusiasm of W. Arber to repeat his discoveries. The test tube in a laboratory is the ideal place to hold cells and extract the cutting enzymes you will need to create recombinant DNA. The test tube is used to cut any DNA you want to work with. The DNA fragment that comes from a cut made by the enzyme can be used to splice together another DNA fragment or molecule, and bingo, you just created recombinant DNA.

If during the cutting and splicing together of small and large fragments of DNA sequences the recipient of the DNA you are using happens to be a chromosome, you will get another. You can introduce it into its host cell, where it will produce new genes. DNA fragments (genes) from any source can be transferred to any cell. We are all made of DNA. It is the variations—mutations—within genes in the DNA molecule that made species grow differently.

There is no room for doubt if you follow the footsteps of these great pioneers in DNA technology. There are many cutting nucleases in nature just waiting for you to begin a career in this field. Making cuts and splicing together all sorts of DNA will keep you engaged and employed forever. In my opinion, there is nothing more interesting and amusing than learning to play with your own book of life. You will learn how to engineer a bacterium for use as an antibiotic vector to save sick people and return them to healthy and happy lives. You may become a genetic engineer working through worldwide agencies to multiply crop production by eliminating plagues, saving millions of people from famine and subsequent death.

A Virus Cannot Multiply Alone

You may recall that viruses cannot replicate by their own mechanism. Viruses, those nasty bugs, need to infect a host cell by introducing their DNA. Viruses use the cell reproduction machinery of protein synthesis to multiply themselves. A virus becomes lord and master of the invaded cell. In a short while, there will not be a thing left of the host cell; viruses will be pouring out, looking for other victims to invade and destroy. During previous chapters of this essay, you learned how some cells have developed efficient tools for self-protection and for killing toxic invaders.

I am grateful to many dedicated scientists who pursued their research goals to the end. I have named a few of them for you. For their contributions to science, health, and humanity in general, I have no words to thank them. There are thousands of ex-patients around the world today who have benefited from DNA technology. There is a widely known and dedicated professor of biology, Dr. David Sadava, whose lectures I cannot ignore here. I love to hear him repeat things over and over again while taking real stories from daily life episodes to illustrate and elucidate the message of his lesson. During one of his lectures, he described two methods to get DNA into host cells. I will describe only one. A chromosome is isolated from a host organism and put into a test tube; it is then cut with a restriction enzyme, and the new DNA is spliced into it. "This rDNA is then put into the host cell, and since it is part of a normal chromosome, the host cell thinks that it belongs in the cell, and the rDNA stays there and is replicated."[68]

---

[68] David Sadava, *Understanding Genetics: DNA, Genes, and Their Real-World Applications* (Chantilly, VA: The Teaching Co., 2008).

Illness under the Electron Microscope

THERE ARE MANY ILLNESSES whose etiologies are just beginning to become known under the eye of the electron microscope and related medical research tools. Our immune system has been under the watchful eye of scientists for many years. There are scientists dedicated to finding solutions for our medical illnesses working 24-7. It has become a worldwide endeavor, connecting scientists from all over the world. They are trying to unlock the door to our immune system and its failures resulting in diseases. Among them we have lupus erythematosus. This illness is closely related to our autoimmune system. It causes chronic inflammation that may involve the heart, lungs, skin, joints, kidneys, central nervous system, and circulatory system. Scientists in the field suspect that apoptosis is involved in the etiology of this disease. However, the abnormal production of antigens by B cells has been claimed by outstanding researchers to be the primary culprit of this illness. Like schizophrenia, autism, and bipolar disorder, among other such issues, it presents itself as a syndrome, making it more difficult to follow its pathways in billions of cells. I have seen young people almost crippled by this disease. They become unable to walk and are limited to a wheelchair for life. Their joints, knees, and fingers are deformed by protrusion, making it difficult for them to hold a eating utensils in their hands. This sad condition is followed by depression and occasional suicidal ideations and suicide attempts. I have seen young men and women praying God to take them away because they cannot tolerate their pain anymore.

Anxiety

My first encounter with anxiety that I am aware of took place while attending junior high school. I had a punishing teacher whom I will refer to here as Mr. Delgo. He is the subject of one of my most recent books, titled *An Episodic Toxic Memory*. I did not have to see him or hear his voice to have my heart beat so fast and loud that even my friends used to place their hands on my chest to compare my heartbeat with theirs. He had punished and embarrassed me in front of the classroom. Consequently, I began to experience very embarrassing episodes like wetting my pants, vomiting, having diarrhea, and suffering headaches. Mr. Delgo, my obnoxious teacher, was an ogre occupying many neurons in my brain. I can say that in some respects he stopped me from growing up emotionally and cognitively.

I had Mr. Delgo as my teacher for three very long years. The numerous neurons and synapses involved in this episodic toxic memory were overly active and super-sensitized, and they were firing and releasing neurotransmitters. Ultimately, the super-sensitized neurons produced new dendrite sprouts. These dendrite sprouts brought about neuronal structural and functional changes. A few particular brain regions were the primary recipients and reactive objects of Mr. Delgo's abusive behavior toward me. Among those cells there is are two very old groups of cells on the sides of the temporal lobe that are called amygdalae. The amygdalae are located anterior of another group of old brain cells named the hippocampus. To people in Greece and Rome who were curious and interested in brain anatomy, this primitive brain structure looked like a seahorse, and they named it accordingly.

My first investigation about fear and anxiety was centered on Sigmund Freud's psychoanalytic theory. My oldest brother was a fan of Freud during the late 1940s and early 1950s. I recall I had begun to have nausea, headaches, sleeping problems, and nightmares, aside from my racing heart, which I occasionally compared with that of my father's favorite horse, Canelo. I tried over-the-counter medication and herbal teas. My mother was believed to be an expert botanist among her immediate family and friends. In all honesty, neither my mother's expertise nor OTC medication were of much help. My problem was not in my stomach, despite my nausea and occasional vomiting. My problem resided inside my head, somewhere in an outgrowth of synaptic connections I could not even imagine existed at all. Eric Kandel, Jeanette Norden, and Nancy Andreasen, three of my earliest icons and intellectual stars, helped me get insight into the monster I had right inside my brain, torturing me long after I had left Mr. Delgo behind. I had left behind Delgo in bone, blood, and flesh, but his image, his bullying and all his wicked behavior were inside my brain. Later on, when I moved to live in a great urban area with many experts in brain and mind problems, I spent almost three years on a couch, followed by another three and half years at a psychoanalytic-oriented institute. The institute had undergone significant modification from traditional Freudian theory. I must add that group therapy and group support were extremely helpful.

An Almond-Shaped Cluster of Brain Cells

The almond-shaped brain structure the title of this section refers to is the amygdala, which, along with the hippocampus, is involved in memory and learning. Without that almond-shaped structure positioned anterior to the hippocampus, humans probably would not make it in the jungle. I am not speaking here strictly of the actual jungle. New predators drive expensive cars, wear expensive clothes, make use of most modern cell phones, and hire less-sophisticated predators to do their jobs on street corners or expensive and selective bars.

I like to travel, and on my journeys my money has at times vanished from my pocket, leaving me without any idea about who the thief was. You may argue that the amygdala is losing its usefulness, but perhaps it is the predator that has become smarter. The amygdala acts independently of the nearby cluster of brain cells. The limbic system and prefrontal cortex are emotional power centers. However, the amygdala does not wait for feedback or response from them when it needs to put you in a safe place in a fraction of a second. Otherwise, you would become lunch for a hungry predator. This amygdala of yours is often referred to as the brain's door to emotions.

There is a difference between fear and anxiety. In fear there is an object you are afraid of. In my case, I began to fear Mr. Delgo while I was in his classroom. Later on, when I had moved to live miles away, fear was converted into anxiety. Anxiety has to be defined as a sustained level of elevated apprehension in the absence of an immediate threat. I had moved sixteen hundred miles away from my teacher, but the ogre in my brain kept torturing me day and night. He could not be an immediate threat to me at that distance, but he had already provoked a structural change in my neurons. When I saw a man who looked or spoke like him, anxiety overcame my apparently peaceful mind. When I was enjoying an ice-cream sugar cone and a Mr. Delgo look-alike passed by me, the ice cream turned sour and occasionally made me throw up.

Anxiety is considered by many clinicians to be the most common psychiatric disorder. It may come from different sources, and it exists in all cultures and countries around the world. It may go unrecognized, misdiagnosed, and untreated for many years in all levels of our society. Denial and the traditional stigma attached to individuals receiving help or treatment for brain problems keep people from receiving help. Unfortunately, they remain in hiding rather than getting professional help and regaining their happiness. It goes without saying that alcoholism, drug abuse, marital discord, and abuse have their bases in anxiety.

The amygdala has been implicated in anxiety etiology. Its implication is logical and is based on many laboratory experiments. Chronic stress can kill many cells in the amygdala, thus contributing to anxiety. Dead cells in the amygdala make one forgetful; learning becomes slower, and you become irritated. One's blood pressure goes up, headaches occur, and strokes and heart problems may arise. Most medical doctor visits cause some amount of anxiety in patients.

My story on the amygdala and anxiety is not intended to frighten you. On the contrary, my good intention is to help you understand your emotions. It is best to familiarize yourself with relatively minor problems that can be resolved before they reach the brain and make it their permanent home. Despite the high prevalence of anxiety disorders in all levels of society, the underlying neural circuitry is incompletely understood, according to a Stanford University professor and researcher. Psychotherapy, including cognitive behavioral therapy, gestalt psychology, rational emotive therapy, and goal-specific and action-oriented therapeutic modalities, has been very helpful in dealing with anxiety disorders and depression. There are prescription drugs, such as benzodiazepines, but their side effects and potential for addiction need to be seriously considered before one begins taking them.

In addition to incomplete studies on the amygdala, this investigation deficit can be partially attributed to the complex internal composition of the amygdala itself. The leading author at Stanford, Kay M. Tye, wrote that the amygdala is composed of functionally and morphologically heterogeneous subnuclei with complex interconnectivity. The primary neurotransmitters are glutamate and GABAergic medium spiny neurons. The centromedial amygdala consists of 95 percent GABAergic medium spiny neurons, while the basolateral amygdala is glutamatergic. No conclusive evidence of a specific causal factor or locus for general anxiety at the amygdala was found. Their investigation concluded that anxiety is continuously regulated by balanced antagonistic pathways within the amygdala.[69]

During this essay, as well as previous ones, I have paid particular attention to the amygdala and the hippocampus not only because of their antiquity in the brain but also because they are always on the hot seat. Both are primary survival-oriented clusters of brain cells that have helped this lad reach this point in life's evolution throughout thousands of years. Soon we will identify a particular protein in the amygdala responsible for breaking the established normal balance between glutamate and GABA production, thus provoking an emotional disorder. It is the amygdala's continuous exposure and overexcitation to stress that governs that particular protein.

Likewise, I picked on translation processes, where RNA is always a big player; these processes are a hot spot for human illness. We should not close up and say "Hasta la vista" until we pick another hot spot: cutting and splicing—particularly, splicing involved in many of our illnesses. Some researchers have begun to explore that sensitive area of cell manufacturing machinery. It would be wonderful for all of us if we could address this health issue with embryonic stem cell lines and regenerate a decaying amygdala. It does not escape my mind that progress we have made with an engineered bacterium can be an excellent vector to make appropriate corrections and adjustments in a dysfunctional amygdala.

---

[69] *Nature* 471 (March 17, 2011): 358–62.

An Animal Cell and Amino Acids in Pictures

Following is a table showing the twenty amino acids and stop codons. I arranged it from the largest to the smallest number of codons to code for protein synthesis, responding to my curiosity and convenience. No relationship was intended between the amino acids based on their numbers of codons.

Amino Acids	Codons					
Arginine	AGA	AGG	CGA	CGC	CGG	CGU
Serine	UCA	AGU	AGC	UCC	UCG	-UCU
Leucine	UUA	UUG	CUA	CUG	CUC	CUU
Threonine	ACA	ACC	ACG	ACU		
Valine	GUA	GUC	GUG	GUU		
Glycine	GGA	GGC	GGU	GGG		
Alanine	GCA	GCC	GCG	GCU		
Proline	CCA	CCC	CCG	CCU		
Isoleucine	AUA	AUC	AUU			
Cysteine	UGC	UGU				
Phenylalanine	UUC	UUU				
Tyrosine	UAC	UAU				
Histidine	CAC	CAU				
Glutamic Acid	GAA	GAG				
Aspartic Acid	GAC	GAU				
Asparagine	AAC	AAU				
Glutamine	CAA	CAG				
Lysine	AAA	AAG				
Methionine	AUG					
Tryptophan	UGG					
Stop Codons	UAA	UAG	UGA			

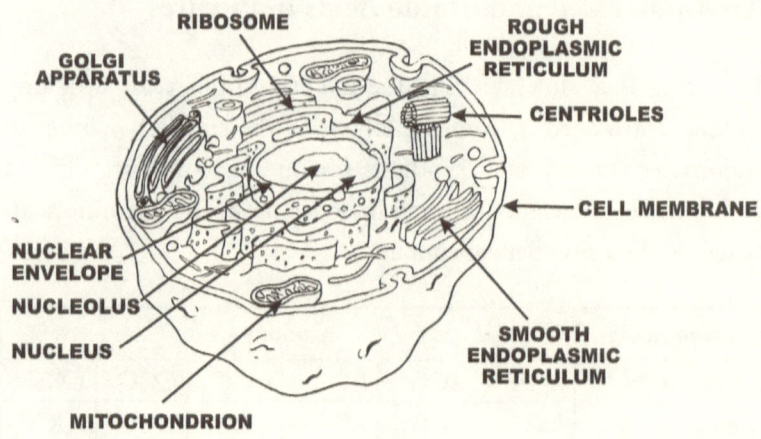

Inside an Animal Cell

There are dynamic chemical and molecular processes taking place inside each cell during every fraction of each second of our lives. Most of these processes involve extremely tiny units that can be seen only under an electron microscope. Equally small are multiple organelles that travel back and forth inside the cell plasma, carrying and delivering the organic material necessary for the body. Some of these are the endoplasmic reticulum, the Golgi apparatus, and lysosomes..

A tiny portion of the plasma membrane on the surface of the cell may separate and cut itself from the membrane, capture necessary neighboring subunits, and bring them inside the cell for processing. Similarly, these tiny cut-off portions may be used as carriers to nearby cells. There are always intercellular captures of material occurring for cell processing and use. Captured external material is usually fused with lysosomes. These cellular vessels function as degenerative machines, digesting everything that passes by. Part of this dynamic process is importing and exporting cell material.

The nucleus of the cell is surrounded and protected by a

membrane often called the nucleus envelope. Right next to the nucleus is the endoplasmic reticulum, which is always busy processing and packing material inside the cell. Within this particular cell vesicle, you will find the protein assembly factory—the ribosome. In each eukaryote cell, meaning cells with a nucleus, there are many compartments with specific roles. The aqueous material in the cell—a gel-like substance—is called cytosol. Besides serving as a cushion, many cellular actions and reactions take place in the cytosol. I must remind you that the energy-producing organelle, the mitochondrion, resides in the cell as well. Also present is the nucleolus; you may call this RNA's home. Through the Internet, you will be able to find beautiful and accurate images of the brain. Look for the groups of cells inside the brain that I have mentioned in previous sections.

From Research to Clinical Practice

It was a hot and humid mid-July evening in Brooklyn, New York, in 2011. An earthquake and Hurricane Irene had recently made unwelcome visits to us. I had already taken a trip around the world in the Internet searching for outstanding news and discoveries in science, economics, and politics that could have an impact on my life. For scientific discoveries, I look for and rely upon the most prestigious journals and books published around the world. My main concerns about world politics and economics are the Middle East, Russia, Europe, and of course, the world's biggest factory, China. It is not that I ignore Africa and South America—especially Brazil. This huge country with immense resources and recently discovered offshore oil deposits will influence world economics and politics in a fashion unpredicted a few years ago. China and Russia are opening doors in South America that may soon lead to it becoming the most competitive world region besides the Far East.

The things transforming the sleeping giant into a world power with its own particular interest and agenda are not just agricultural genetic engineering, lumber, and beef production but also homegrown and imported scientific technology. These issues keep my brain active in making relevant associations and connections between dendrites. This mental exercise helps strengthen synapses, making new memories that may become permanent in my brain. I do not ignore the old saying "Use it or lose it." It keeps my brain from becoming rusty, just as my grandfather used to say.

At home, my bedroom is full of books and journals that are mainly on neuroscience, psychology, and genetics. I do not do much exercise because of my breathing problems; I developed emphysema. I did not order emphysema online; it just came into my lungs without an invitation. I take medication on a regular basis. I walk a few blocks every day, and I am choosy as to what I eat and drink. I call myself a happy and blessed person. People stop me for all types of questions regarding health and world politics. My engagement with them is not in the form of a linear or one-way dialogue but takes many directions; I often answer a question with another question. Many times I tell them that if I answer their questions, I will deprive them of the joy of finding the truth by themselves. This is done based on my conviction after many years of clinical practice. Repeating somebody else's answers, whether true or false, is robot-like behavior.

I am not saying that you do not need to learn from your neighbor. What I am saying is that each one of us is a unique individual unlike anybody else in the universe. Whatever truth you are seeking out, it is your own truth, not your neighbor's truth. Even identical twins do not share the same truth. Each one of us has a genetic code containing information many hundreds of years old. Besides, molecular and chemical reactions taking place, above all, in our brains, makes each of us different from anybody else here and now. The brain avoids many problems and diseases if it is busy working on something that will benefit us all. My daily chats with people are a form of socialization that is helpful to me and others. There are people twenty or more years younger than I am who keep asking me for my recipe, meaning my daily diet. Besides my local chats, I keep in contact with friends and colleagues through the Internet or via cell phone. After I retired from my clinical practice after my seventy-ninth birthday, I have had a bad habit of visiting my friends at home or at their offices. They have all responded happily to my visits. I have written all this for you because I want to share with you a recent unhappy experience that has been bugging my brain, and I need your input.

On that hot and humid mid-July evening, it was 11:00 p.m. when I decided to call it a day and go to sleep. I confess that I have a bad sleeping habit of grabbing a book, journal, or magazine for my last light reading for the night before I turn off the lights in my room by pulling a yellow cord located next to my left side of my head. Believe it or not, the purpose of this last reading is to relax me into sleep. I purposely pick up something that will not be engaging to my brain. I do not want to read of a scientific discovery, a new medical discovery, or the latest DNA technology that late at night, as that may excite my brain cells into further investigation. I have to spend my last few minutes before my brain's sleep-wake circuit is turned off reading innocent and soothing stories. Unfortunately I do not find those pleasant and sleep-inducing pages all the time. And this hot mid-July night was one unfortunate and miserable experience for me. Do not get me wrong, please. This is not the only painful and tortuous night I have experienced. I have had more than just a share of these during my life time. However, most of the time, these experiences serve me well as good learning opportunities.

I'll bet that either you or I will spend eighty-three years of life on planet Earth just listening to old, sweet children's stories. I love those stories, but there is a time and place for them. I was not born in a golden crib, and I attended overcrowded public schools. I am just like hundreds of millions of children, adolescents, and adults who grew up with heat, hunger, and unjustified rejection and punishment. These things become integral parts of our lifelong trip on this beautiful home planet of ours. I am not implying that growing up is an easy task for a human embryo or any other organism, but for many of us, the rules of nature have been heavy-handed.

But enough of poverty, disease, hunger, and cruelty of all sorts. The world's standard of living for human beings has never been as good as it is now. I myself have had pleasant, beautiful, and educational experiences that make me greatly appreciate the beauty and goodness of this planet and its people. I try to enjoy it as much as possible.

There was hardly any cluster of brain cells that did not become active and provoke unwanted behavior in me, though I was not aware of this for many, many years. My brain has a habit of going traveling in space and time, even against my will. It takes me to places and times I have never imagined existed even in fairy tales. And believe me, I have read many fairy tales, beginning with *Don Quijote de la Mancha*.

Before I forget it, I am going to tell you a secret I did not reveal even to my mother. I Once dreamed I could write a book like *Don Quijote*. This is not a joke; I dreamed such nonsense. Go ahead and have a good laugh at my expense, but it is true, I had such contempt for Cervantes that even unconsciously, while dreaming, I made fun of him.

I am now going to cut short this preamble to the experience that drove me almost insane on that otherwise quiet and peaceful night. You can call me a bleeding heart, a coward, a panicky old man, or a messenger from the heavens, and I will forgive you. Forgiveness, as I see it, is a powerful healing tool. Anger and jealousy, on the other hand, are as poisonous to your health as a cancer that eats away cell after cell before it is discovered. We have the tendency to deny our anger while it destroys us inside. Our brains suffer, our hearts suffer, and our stomachs develop ulcers if we do not become honest with ourselves. We can spend a lifetime defending our own lies. We behave even worse when we buy all types of remedies to cover up our ulcers and headaches. If you allow your organism to take care itself without your unnecessary interference, you will be better off.

I have to implore your mercy in asking that you do not too harshly condemn my emotional response that horrible evening. The motive that triggered an unanticipated, spontaneous, unconscious, and automatic affective response on me that night made me revisit my childhood life experience in its entirety. Be prepared for what is coming; do not hold back your emotional response.

Remember: the first thing a baby does when he or she comes out of the comfort inside the mother's womb is cry. The visual trigger that provoked my emotional state was a girl about six or seven years old who had been chained to a tree trunk with a strap above her left ankle. I repeat, the girl was chained to a tree trunk. The image was shown in large scale at the beginning of a highly professional and prestigious scientific journal. Please wait to get the whole picture before making any judgment. My intention that night, although naive, was to get a restful and enjoyable full night of sleep. That had always been my mother's last wish for me at night before I plunged into deep sleep.

Right below the picture, the article's author explained that the girl had been punished by her family and neighbors because she claimed to have seen and talked with people not visible to others. It is hard for me to believe that there is anyone who has not heard a story like this one before. In our daily parlance, the girl was having visual and auditory hallucinations. Some cultures may claim the girl was blessed from heavens for the good of the many. However, in this case, the girl was punished because her family and her community believed she was bewitched by evil spirits. They believed she would bring diseases and death to her neighbors. The picture and the story about this young girl were published for public view during the year 2011, not four or five hundred years ago. It was taking place in our time and on our planet. It was not a television show or a documentary aimed at a specific audience for a commercial purpose.

Incidentally, on July 18, 2011, a public television channel aired a documentary about a southeastern Asian country where schizophrenic patients were kept in bamboo cages. They were considered a threat to society. A shaman, not a mental health professional, was responsible for the diagnosis and treatment of brain problems and diseases. This brain disease is not limited to a country, race, or social or economic class; but diagnosis and treatment in underdeveloped countries is extremely poor and inadequate. Industrialized and relatively rich Western countries have well-trained professionals and medicine available to control most of the symptoms of this disease. Most people on planet Earth do not know that genes and biochemical markers in brain cells may be responsible for the voices the poor girl was experiencing.

I am still trying to break down my behavior during that night into its component pieces for proper analysis and digestion. What is to be assimilated, if it needs to be, it is too early to tell. Before I could place the journal and the picture aside, I noticed that tears were running down both of my cheeks. Also, my nose was getting wet. I have worked with children, adolescents, and adults for many years. My experience at a psychiatric emergency room placed me in contact with patients whose problems broke down even the best-trained highly experienced therapists. Whether or not you believe it, the pain of others will sooner or later begin to wear you down. It becomes a revolving door; therapists left the job no matter how comfortable we tried to make it for them.

Yes, on that hot night, I found myself crying for a little girl I had never seen. Whether the tears were for the chained little girl needs to be further clarified, explored, and analyzed. We carry emotional scars in the form of toxic episodic memories that may be triggered by the pain of other human beings. Even the abuse of kittens, puppies, chicks, or any other defenseless animal may serve as an emotional trigger for a physical response.

My eyes saw a girl in pain tied up to a tree trunk; the message was sent to my visual cortex in the back of my head. From here it must have gone to the thalamus—a relay station composed of many complicated brain cells deep in the middle of my brain. The limbic system on and around the thalamus was immediately involved, including the prefrontal cortex. Groups of brain cells included in the limbic system seem to be an arbitrary one. There is not a consensus among neuro scientists as to which brain cell nuclei should be included in this system, but most agree that it is a very primitive survival region of the brain and the seat of emotions. And when we talk about survival, the amygdala takes a front seat. The amygdala has helped me survive my eighty-five years of age; it holds within itself automatic responses from thousands of years of human growth and development. I will come back to the limbic system later on in this essay. I hope I have already named most clusters of neurons included in this system.

I am careful not to contaminate my brain with dangerous drugs, toxic memories, or grudges against anyone. When you hold a grudge against another human being, you are hurting yourself. You have turned anger inward, and it manifests itself in many bodily symptoms you try to cover up. You may drink or swallow all sorts of things to calm down your symptoms, but in reality you are simply hiding your lies from yourself.

For my tears to have flowed freely despite my conscious awareness, they must have been triggered by deep brain pain spots that might have been formed years prior. Some of my old experiences have become permanent memories stored in my brain in a place I cannot identify at present. But I do know that the hippocampus is involved in retrieving past memories.

You must be familiar with the accidental case of Henry Molaison, often referred to simply as H. M., who could not remember anything after surgical removal of both hippocampi from his temporal lobes. The tragic case of H. M. and most recent studies have confirmed without any doubt the function of the hippocampus in forming and retrieving memories. Anterior to, and very close to, the hippocampus, we find the amygdala, which is strongly involved in learned fear. The big tragedy in H. M.'s case was that both hippocampi were removed. Each temporal lobe has one. H. M. suffered from chronic and very painful epileptic seizures.

PET scanning and functional fMRI, two sophisticated technologies, have been very helpful during the diagnosis of brain diseases. The use of in vivo imaging techniques while working with patients in clinics and hospitals gives scientists a very clear and accurate picture; an abnormality in structure or function, or damage to a brain cluster of cells, will be obvious to the expert eye of the scientist doing the job.

I used to say that the picture struck me like lightning, but in retrospect I must modify the term, because I have not seen or heard of anybody surviving such a punishment from the sky. Besides, I am trying to find connections between the lightning, the metal chain, the girl, and my bed. As a country boy, the first thing that comes to my mind is lightning hitting palm trees and avocado trees and splitting them into pieces. That is a very frightening and dangerous experience for anyone, young or old. I will leave lightning alone and address my fear and the tears on my cheeks.

The first sight of the girl looking down, apparently in tears, must have struck a group of my brain cells containing very old and painful memories from at least seventy-five years before. When I was about seven years old, I saw an old lady with her hands and feet tied together taken to a hospital because she had set fire to her family home. Later on, as an adult, I learned she had a schizophrenic daughter and a very violent son. During my private practice, I worked with a patient whose depressed mother tried to kill him at age five by tying him to a tree trunk and setting fire to a nearby ranch. There are painful memories in my brain that could have been triggered to action by the sight of a chained little girl. However, it seems that my first unconscious protective response was to ignore what I had seen and turn to the next page.

What I had already seen, but did not want to see, must have been associated with hidden memories repressed by defensive layers of denial. Denial and repression may help you momentarily, but in the long run they will catch up and get you when least expected. When I had already turned my attention to the next page, trying to deny what I had already seen, a slight feeling of anger showed up in me. Quickly, and without hesitation, I went back to the little girl in chains I was trying to ignore. I turned on a second light in my room, and it convinced me that my brain was not playing tricks on me through unconnected designs and visual patterns.

Not totally convinced that I was seeing something taking place on our beautiful planet during the twenty-first century, I took another look at the cover page. My brain was not playing tricks on me; the date on the front of the page read "July 2011." I went back to the page with the chained girl and looked very closely at the chain. I was looking for something to soften the girl's pain—which, in reality, was my pain. I was trying to change the metal chain into something less painful and ugly to my eyes. I was trying to foist humanity upon the beast that had chained up an innocent human being, to assuage his or her own ignorance and wicked behavior in my eyes. No matter how hard I tried, the beast was overpowering, and the chain got heavier and more brutal every time I looked at it. I saw men, women, and children in chains all over the world. I could not help it, but my brain took over and sent me through time and space. I mentally travelled to Egypt, Persia, Babylon, Greece, Rome, Africa, and America. These things are still taking place in the twenty-first century. We punish people for things we do not understand.

I saw nothing but chains covered in the blood of innocent men, women, and children. It looked as if a bestial creature were feeding on human suffering and blood. However, after I attempted to uncover the beast, what I found was man's thirst for power and greed. The beast is a man-made animal; it enjoys torturing and killing other human beings. The main focus had been the girl who was suffering in pain because she had shared with family members and friends things she alone had seen and heard. She might have been asking for support among trusting family members. My brain tried to soften my own pain by projecting pain on other human beings and took me on a mental world trip. I have to be grateful and thank my brain for trying to ease my pain; it has done so for many years. My experience may also be interpreted as an unconscious attempt to find an excuse, a rationalization, to continue to deny the little girl had pushed pain buttons in my brain. Perhaps it was true that the chained little girl had exposed my weakness.

Despite my intense training and clinical experience, there are scars that are not completely healed. My attention was divided between two different objects: the little girl and the chain. When I realized my brain was easing my hidden pain by projecting pain upon many people around the world, I went back to the innocent and defenseless victim. It was then that I could see, almost hidden between different shades of color in the picture, tears coming out of her eyes. You must remember that I was not aware I was crying. My tears were running down my cheeks, and my nose was getting wet, but I kept looking at the picture as if it were incomplete. I kept looking at the metal chain, wondering if it could have been made of a softer material. A soft material could perhaps have made the pain more tolerable. Actually, I had taken my eyes off the triggering stimulus that had provoked my own tears: the little girl's tears. My mental world trip could have softened my own pain and put me to sleep, thus preventing me from going back to the visual trigger that had provoked my silent tears. My brain wanted to spare me from further pain and chose the chain as an attractive object to focus my attention upon.

Tears by themselves can be triggered by various stimuli. Some people shed tears of joy when they meet family members or receive good news from neighbors and relatives. I have seen men and women shed tears when their baseball team won a championship. And I have seen people cry over the death of a puppy or kitten. The determination of the stimulus that triggers the shedding of tears seems to be a complicated issue.

Professor Antonio Damasio is head of the Department of Neurology at Iowa University Medical Center. In his book *Looking for Spinoza*, he describes an extremely interesting episode of tear-shedding that relates to my case. It is a documents the case of a sixty-seven-year-old lady with a history of Parkinson's disease who stopped responding to conventional medical treatment. She was a candidate for deep brain stimulation, which required the implantation of two tiny electrodes in her brain stem. The brain stem is a primitive region of the brain that begins in the upper part of the spinal cord. Generally, brain anatomists divide the brain stem into three sections: medulla, pons, and midbrain. It is a survival-oriented region of the brain because it controls many vital organs of the body.

Surgery to implant the electrodes took place in a hospital in Paris, France. The patient, Jeanne (not her real name), had no history of depression; nor did anyone in her immediate family. She was one of about nineteen patients who underwent the implantation procedure at the hospital with the same surgical team. Each small electrode has tiny points of contact that can deliver a controlled electrical current. Unfortunately, in Jeanne the electrical current missed the target by a tiny fraction of two millimeters and touched what Dr. Damasio called an emotional button that was linked to sadness. Otherwise, she responded extremely well to the surgery. To the surprise of the medical team, Jeanne began to cry, shedding tears and sobbing loudly. She even verbalized to the staff that she wanted to terminate her life, meaning she did not want to live anymore. She was calling death to come and take her away. She had gone in for surgery with the intention of living and enjoying a few more years on this beautiful home planet of ours. However, a little mistake in her brain changed her mood from happiness to severe depression. The electric current turned on a trigger point in her brain that provoked a depressive state and suicidal ideations. There was no history of depression in Jeanne's life history. As far as the medical team was aware, Jeanne was a happy-go-lucky person, except for her bout with Parkinson's disease.

I brought this case to you because I have been talking of pushing pain buttons we are not aware of. This is another confirmation of the existence of pain buttons in our brains. In this case, the action of pushing the button on Jeanne was done by an electric current. Jeanne was not complaining of having pain because the electrodes were hurting her; the brain itself has no pain receptor cells. Your body's skin from head to toe is responsible for those receptors.

After the electrode position was corrected, Jeanne gave not a single sob; nor did she shed tears. Her death wish disappeared from her like magic. She became the same jovial, talkative, and happy lady as she had been before her surgery. An interesting observation Dr. Damasio does not want to go unnoticed by us is that the sadness came first and feelings—a subjective state—came afterward.

My case with the little girl shedding tears while chained to a tree trunk is still unresolved. I cannot blame electrodes or anything else in a surgical procedure as the causal agent. I was alone and comfortably lying down on my bed. The messengers that carried the information that provoked my tears and, consequently, the feeling of guilt and anger were my two eyes. How the message went to a specific pain button that late in the evening when I was getting ready to set aside the day's challenges and go to sleep is another question to answer. I had eaten a good dinner at around 6:00 p.m., and I was rested before I decided to say good night to everything around me. Unlike Jeanne, I did not have anyone working on my brain or asking me provocative questions. I can say that I was happy that evening; no intrusive thoughts were breaking the harmonious chemical balance in my brain.

Well, I have to admit that I cried over a picture of a girl I had never seen before. But there is another angle to this picture I have not addressed it yet. This one is not an intellectual exercise, but it is something that sits at the base of my emotional response to this painful experience. Why did I respond with anger when I realized I was running away from the real issue at hand? Why did I experience anger, and where did it come from? Why did tears and anger come together? How is the little boy in me involved in these emotions? Do I have to go back to my childhood life experience to try to explain my adult tears and anger over a picture of a little girl crying while tied up to a tree trunk? I can make use of the Freudian psychoanalytic theory to try to explain, both my tears and my anger. I can look to the ego as the main agent delivering messages and mediating psychic conflicts between my id and superego. However, I prefer to stay in the twenty-first century and look to neuroscience to explain my behavior based on scientific proven models.

There appear to be significant differences in life events between Jeanne and me. She did not suffer from depression and was a happy lady all her life. I, for one, did not enjoy a happy childhood. My mother had eleven children, and my brother born right before me, suffered from what appeared to be thalassemia major and died when he was a little over two years old. My mother had delegated my care to my oldest sister. Secondly, while I was attending intermediate school—meaning the seventh, eighth, and ninth grades—a teacher named Mr. Delgo forced a very painful, traumatic, and toxic memory on me. See my book *An Episodic Toxic Memory* for details. I spent years in therapy, trying to bring closure to a dysthymic disorder. It could be called a mild version of depression. Psychoanalytic-oriented therapists would not hesitate to connect my tears to my childhood experience, but I prefer to look to modern science and technology.

In Jeanne's case, Dr. Damasio does not tell us the specific loci in the brain stem where the electrodes that provoked tears and sadness were implanted. That section of the brain contains many clusters of brain cells responsible for myriad behavioral responses. Besides, the brain stem appears to be similar to a superhighway that is extremely busy, with messages going up and down it. Messages from the body go up the spinal cord to the cortex, and likewise, the cortex sends messages down to the rest of the organism through the brain stem. Adding to the complexity of this region of the brain, it contains clusters of cells responsible for generating chemical messages like serotonin and norepinephrine; these messages are responsible for behavior and memory formation. There is another group of cells around the medial superior part of the pons, not far from the locus coeruleus and Raphe nuclei, known as the substantia nigra compacta. When its cells die out, it no longer produces dopamine. When there is a lack of dopamine from this brain nucleus, the brain's owner develops Parkinson's disease. This is the illness Jeanne was suffering from, and the electrodes were implanted to make up for the deficit of dopamine. The dopamine released from the substantia nigra compacta goes to two nearby cell nuclei known as caudate and putamen. Dopamine is involved not only in Parkinson's disease but also in schizophrenia.

I am not trying to get you involved in brain anatomy and physiology. It would be wonderful if you were to do so, but what I am trying to say is that it is not easy for me to say, "Here is the solution of the problem for you." People laugh, cry, and shed tears for many reasons they themselves cannot explain. My first reaction was to run away from the girl's tears, but I returned to her in anger. Anatomically, the sites related to tears and anger are not far from each other. Tears come out of your eyes and are visible to an observer, but the location that triggers them was visible only to Jeanne's surgeon through sophisticated scanning machines.

Anger, another emotion, has many ways of manifesting itself, but the anger-related locus in the brain is not far from the locus related to tears. The hypothalamus—deep in the center of the brain, but not part of the neocortex, which developed later on—is involved in anger genesis, among many other functions. The hypothalamus and the pituitary gland, a crucial endocrine gland, are structurally and functionally like identical twin brothers; their behavior is inseparable. There are hormones that travel from one cell nucleus to another carrying messages that can be translated into anger.

For the sake of clarification, let's say that Pete responds with anger to Mike's joke, while John just laughs at it. Mike might have taken a stone and hit a stray dog on the street. Pete might have passed from an anger phase and engaged in a fight with Mike over the injured dog. In this hypothetical case, there are two potential sources of anger: a visual source and an auditory source. John remained neutral. Pete's brain responded to two old and powerful forms of sensory input. You may point out that Pete's parents' religion and his cultural background must have contributed to his behavior. Pete's upbringing might have been different from Mike's and John's. Pete also may have been responding to Mike's bullying and aggressive behavior. Mike hitting a stray dog burst opened the anger bubble. A section of Pete's brain accumulated painful experiences and feelings until it could no longer tolerate them.

In Jeanne's case, her tears and suicidal ideations were not provoked by religion, culture, or a car accident. A piece of metal had mistakenly touched a tiny piece of brain tissue. Instead of touching an anger button, the metal came in contact with a depression-provoking locus.

Tears and anger have multiple functions. Tears are a lubricant that cleans and protects our eyes. Anger, besides its function in aggression, can thrust you into battle in the defense of your country and family. During the research section of this essay, I took time to expose myriad cell responses and behaviors. With that in mind, please try to understand how difficult it is for me to explain my emotional response to the little girl in chains.

Returning to my experience, unlike Jeanne's and Pete's cases, there were no electrodes or bullies to provoke the shedding of tears. It was a picture of an innocent little girl chained to a tree trunk. She was not hurting anyone. She was attempting to get help from family and neighbors to explain something she could not understand. In Jeanne's case, a professional health team was helping her overcome a dopamine deficit and reestablish normality in her life. I am assuming that Jeanne was neither verbally assaulted nor heard anyone screaming while she was undergoing surgery. In other words, she was not receiving painful visual and auditory stimuli. She did not have Mike to bully her.

In my case, I was bullied during childhood and as adolescent too. Can these past experiences from over sixty-five years ago have anything to do with my shedding tears? If I am not mistaken, most people would dismiss this as nonsense. They would argue that these experiences must have been forgotten long ago. It seems logical to most people that time would delete past bad memories. Our culture encourages us to ignore childhood experiences or play down their significance during adult life. We are told the same thing over and over again by peers and family members. People around us often say things like "You are man [or a lady] now; forget childhood stories. We are adults now; we should not be entertaining unimportant nuisances from earlier years." Boys are encouraged to avoid crying, while girls can shed as many tears as they want to.

Dr. Antonio Damasio rightfully reminds people that they are emotional beings. During my own clinical experience, I dealt with emotions in many details. I believe that it is a strong point to attribute anger as well as the emotion of sadness to a biological base in the case of Jeanne. The fact that she responded to a touch of a tiny piece of metal by shedding tears and crying without any other provocation amply supports my thesis: humans are basically emotional beings. We are conditioned to repress feelings at the expense of our mental and physical health.

On each side of our brain, in the temporal lobes, we have a master nucleus of emotions called the amygdala. This very old cluster of brain cells has helped us survive many years of jungle life. We have carried the jungle with us to urban areas, where violence against each other is a day-to-day survival experience. Of course our amygdalae have made appropriate adjustments to meet the demands of traffic jams, job competition and insecurity, street gangs, and family needs, among many other things.

The dynamics that triggered my tears are at present unresolved. What provoked me to look deeper into my emotional makeup are multiple causal factors. The tears came out spontaneously, as if they were being pushed out by an internal mechanism beyond my conscious control. I was not sobbing or screaming out of pain, but my emotional self was physically responding to a stimulus. It was a visual stimulus that called for a response involving the shedding of tears. Was the emotion of sadness involved in this case? Please remember that Jeanne was shedding tears and verbalizing a desire to finish her life although she had been a jovial and happy patient mere minutes before. She caught the surgery team by surprise and emotionally unprepared to give an appropriate response when she verbalized her wishes to die. I responded to my tears by going turning to the next page of the magazine. I could argue that my unconscious defense system is intact and well prepared to handle unwanted emotional situations and challenges. I can formulate this scenario in this way to facilitate my point. My defense system experienced my pain and ordered my motor cortex responsible for moving my hands and fingers to turn the page. By turning the page, I would not only stop looking at a painful stimulus but would also change to a more pleasant and less provoking stimulus. It might have changed my emotional state at the time. I would have stopped shedding tears. That would be similar to my amygdala telling me, in a case of danger, "Legs, run as fast as you can before a mad bull gores you." Whether this is a poor comparison is a different topic, but I hope you get the point I am trying to get across to you.

When I became aware that I was running away from the challenge the picture of the little girl was posing on me, I became angry—angry at myself. I was not asleep or sedated; I did not turn to the next page unconsciously. My defense system was protecting me from shedding more tears; it was blocking unconscious pain receptors. This attempt to find a solution to my tears being followed by anger is not an uncommon type of response. When I was running from a mad bull on my father's farm and was looking for a safe place, or if I got bruises or was stung by a honeybee, my defense system would take care of it and spare me the pain. After I had gotten to a safe place, the pain would emerge.

You and I have built an organism throughout our evolutionary journey that has learned many tricks to take care itself. Blood clotting is a very useful and protective tool that prevents us from bleeding to death. The feeling of pain is another trick your body developed to let you know there is something wrong going on in your organism. Even inflammation under our skin is a protective device; it means your immune system is working. It means your T cells and legions of its assistant cells are doing their jobs. They are placing the invading pathogens in a corral for the final kill. Even tears have a protective role for your eyes. No, I am not running away from the main question at hand. I am supporting the thesis that both my tears and anger could be defense mechanisms developed by my organism to spare me from additional unnecessary pain.

During Jeanne's surgery, the removal of the electrodes stopped all her symptoms of depression. In my case, my organism's past experience with pain must have put a stop signal in action at a gene or protein level. The visual sensory stimulus had triggered a gene or genes to synthesize proteins that would mobilize a whole mechanism for the shedding of tears. I hope that you agree with me that we are not conscious of actions taken at cellular, molecular, and atomic levels. I do not know what my genes are doing as I type these words. I have a conceptual knowledge about my brain's many actions taken while I think and while I type what I am thinking. However, I am not aware of all the things my brain is accomplishing. Can you guess how many chemical enzyme reactions have been provoked in order for me to write the things I am writing right now? Computers can come up with a good estimate, but it is still an educated guess as to what is really happening inside my brain.

I am not trying to make things more complicated than they already are. Tears and anger provoked by a photograph of a girl I did not know and had never seen before are complicated enough without me adding anything else. First, I have not seen anybody torturing an innocent and defenseless child in that fashion, though I did once read a story of a five-year-old boy who was almost killed by his mother during a psychotic episode. Actually, it was a picture on a piece of paper that provoked both physiological responses on my part. As far as I know, I was not suffering from a disease that might have provoked unwanted tears and anger. I had previously gotten a complete physical examination, including blood work.

For readers with a strong religious orientation, the picture of Christ and Mary may provoke tears. During my visits to Fatima in Portugal and Guadalupe's shrine in Mexico, I have seen many people cry and shed tears. I would be on a completely different route if I pursued my experience under psychodynamics or through religious interpretation. I am trying to see it from the perspective of an organism that is rational, persevering, and emotionally aware of its complexity. This organism believes that emotions are at the base of humanity. It is an attribute among mammals that glues family ties, keeps tribes together, and prompts a person to risk his life for another. Emotions are communication tools that have helped humans build great cities and great cultures.

The little girl in the picture can be considered a communication tool to arouse emotions in another person. I never gave much thought as to whether it was a propaganda trick by the journalist to solicit help and monetary contributions for that country. It would not have helped me any. It would have come after the water was spilt. A visual stimulus had triggered a very complex physiological and behavioral response process that I could not control at a conscious level. At the moment, I feel I must continue pursuing not the motive behind the tears, but rather that of the anger I felt afterward.

Most people are aware of, or at least have heard or read about, the behavioral trilogy: the fight, flight, or freeze response. In the case at hand, a visual stimulus was responsible for provoking my initial response of sadness, which was outwardly manifested by tears and a runny nose. My unconscious must have had connected consonant past painful memories with the pain I imagined the little girl was experiencing. To repeat myself, I had no reason to suspect the journalist was using the little girl as a ploy for his own agenda. Sharing pain with one's neighbor is not unique to humans. Elephants draw close to each other in order to make an impregnable wall when protecting a wounded calf. Honeybees and wild ants—not the ones scientists usually make use of for research in laboratories—make individual sacrifices for the benefit of the collective unit. One may argue that there is a collective tendency at the molecular and cellular level to identify with and rescue a member or a relative of one's genus. Of course, we have to consider, in evolutionary terms, when we parted from each other a long time ago. For instance, take the similarity between the human genome and that of the mouse. Not many people would shed tears spontaneously over a mouse that was killed while trying to eat their expensive cheese. I suppose that you will agree with me that our behavior toward cats and dogs is quite different. Both pets share our food and shelter.

Suppose I proceeded to identify and name all the neural dynamics that were involved and necessary to produce the behavioral and physiological response on my part when I looked at the picture of the girl; it would definitely fall short of explaining my experience. The beauty of this argument, if you allow me to call it an argument, is that it provokes me; it motivates me to continue to explore and search for my identity. It leads me to identify and connect with my old self—my emotional self. This old emotional self must have its place of residence in the limbic system.

For the sake of continuing the argument, the limbic system was developed after the primitive brain stem but before, or simultaneously with, regions and clusters of cells in the neocortex. The prefrontal cortex is considered by many neuroscientists to be part of the limbic system. A very primitive subcortical cluster of brain cells, the amygdala, the master of learned fear and the door to our emotions, is in a constant state of alert. If you add handling chronic stress to its function, its cells will become dysfunctional or die. There is an amygdala in each temporal lobe. You are a very complex organism.

Why should I go in pursuit of anger if I have not yet found out to any degree of satisfaction what triggered my tears? Well, anger is a very powerful emotion not to try to go to the root of its provocation. We humans seem to often get angry at the slightest provocation. It can get us in serious trouble. I would like to equate anger with poison. It not only clouds our ability to think and act rationally and intelligently but may also lead to serious physical and emotional problems, besides souring our social lives. If you are an angry salesman, your family may go hungry and the mortgage payment might end up in jeopardy. In my profession, I cannot afford to be angry. It is imperative that I pursue the trigger that provoked me to get angry at myself for turning to the next page in the magazine and trying to dismiss my tears as just simple nonsense.

Furthermore, tears, sadness, and anger can be construed as symptoms of depression. You may think that I am overextending by linking a seemingly trivial episode with a brain disease. Well, I am going to share with you that a person suffering from severe depression may attempt to kill herself and succeed. She will kill herself when she experiences enough anger toward herself. She might fool the psychotherapist by verbalizing that she is doing better, while in reality she is planning suicide. Now, do not get me wrong, please. I am not equating my experience of anger and tears with severe depression. The thing, if it is a thing, that provoked anger in me must have been a sense of guilt. See, anger, as I said earlier, has its place of residence deep in my brain. It involves, among other groups of brain cells, the hypothalamus, which will provoke a cascade of hormones, mobilizing the entire central nervous system. Do not underestimate the role of the old brain stem, with its locus coeruleus and the Raphe nuclei, in this brain orchestra. The Raphe nuclei, located in the upper posterior section of the pons, right above the medulla, is a rich nucleus of the neurotransmitter serotonin. A deficit of serotonin is related to depression. "Pons" comes from the Latin noun "*pons*" which means "bridge." The locus coeruleus is located near the Raphe nuclei and is a rich area of norepinephrine cells. In general, this is an excitatory chemical of this part of the brain stem. As you may have already guessed, both of these two primitive groups of cells have many connections with the limbic system, including the amygdale and hippocampi. The medulla is an extention of the spinal cord. These are very primitive groups of cells that have existed for a long, long time.

I am not trying to make it hard for you to understand my anger. I am just trying to show you how old these primitive nuclei surviving in our organism are. Quite interestingly, in case you did not see it, those two cell nuclei seem to complement each other. A lack of serotonin from the Raphe nucleus makes you feel down and exhausted, and leaves you in no mood to enjoy your life and your body. The locus coeruleus, on the other hand, makes you feel excited and puts you in a challenging mood. I am not suggesting there is a fighting match between these two cell chemicals, but they act as factors of life conditioning. One may be an attribute useful for hunters and guerrilla fighters, while serotonin may lead to farming, cooking, and household chores. As far as I know, there is no scientific datum to support this hypothesis.

Aside from the biology of depression, there are environmental causal factors that are equally important in the etiology of depression. There are well-documented professional reports that infant and childhood motherly neglect and deprivation are conducive to depression and disease. There is abundant documentation by teachers and psychologist that neglected infants and children do poorly in academic work. They experience sadness, and even their physical growth seems slow. I am are not trying to annoy anyone, but if I am seriously looking for a biological trigger for anger, I must consider, among other things, gene mutations. I also should include epigenetics and environment, diet, illness, and harmful habits like alcohol consumption, smoking, and drug abuse. A poor diet is mostly associated with people in underdeveloped countries. However, the fast-food fad in some rich and developed countries is also a serious dietary problem.

Anger can become a serious health problem because it involves the central nervous system and the peripheral system, with myriad hormones preparing the organism for fight or flight action. This means the heart and the entire circulatory system will be engaged in extra work. Arteries will be constricted, forcing blood to flow faster. The heart will pump blood faster, sending needed blood to the brain. This may cause headaches or even a stroke if one has an elevated level of low-density cholesterol. Needless to say, the heart itself may become a victim of anger. Epinephrine, also known as adrenaline, is a major hormone that provokes many physiological responses. This hormone is produced in response to stress-induced emotions. Anger, anxiety, and fear are a trilogy that seems to have a common denominator. I already mentioned the amygdala as being a master cell nucleus for fear and anxiety.

Fear, anxiety, anger, and aggression can be placed together as survival tools. If we consider anger and aggression as twin emotions that helped us survive in the jungle, we may appreciate their usefulness. During the expression of both emotions, the human body makes use of a lot of energy, facilitated by a powerful and very busy protein—adenosine triphosphate. This protein is engaged in many molecular actions and reactions, even going against the normal flow and behavior of other molecules to achieve a specific objective. ATP consists of one molecule of adenosine to which three phosphate groups are attached. It does not disintegrate easily while performing its catalytic role. Its high-energy phosphate bonds seem to be unique to this hard-working protein. There seems to be hardly any job done in our body that ATP does not donate energy to. ATP releases energy while engaged in its job; therefore, it must replenish itself to resume its expected role of donating energy.

Sugar in its many different forms, like many other nutrients, is an excellent source of energy that must be transported into the cells of all organisms to carry out action. The brain is an organ that consumes large amounts of sugar. Our bodies need a permanent supply of energy for multiple metabolic processes. Our star, the sun, is there for us to make use of it as needed. Our trees and domestic plants, such as potatoes, bananas, beans, and sugar cane, are great sources of energy for our bodies. Also, hydrogen, oxygen, nitrogen, and carbon are elements that exist in great abundance and are available to ATP. When you are running uphill, your inseparable friend ATP is giving you a hand to finish the job.

During its catabolic process—meaning when ATP is releasing energy—ATP's energy may be used to do a job in a cell. When ATP, our never-tiring and generous energy-donating molecule, is helping other molecules do their jobs, we call this process anabolism. While catabolism releases energy, anabolism requires energy. When you kick a ball (anabolism), it goes from a state of inertia into action. When the ball reaches its apex begins to fall (anabolism), it is releasing energy; but when the ball hits your cold glass of soda and spills it, there is another cycle going on. Our bodies are huge complex factories in which of trillions of chemicals reactions take place during every second of our lives. Atoms share and donate electrons; there are many chemical bonds taking place in molecules and cells in our body making use of the good services of ATP. Learn to love your energy-donating macromolecule ATP.

I know you are an intelligent person and may have already thought of a connection existing between ATP and depression. Anger, for one, to be expressed, needs the mobilization of ATP energy in its multiple forms. Resting ions and molecules need a kick to start rolling and get hormones to travel at high speed.

The above-mentioned trilogy of fear, anger, and aggression helped us overcome many challenges in the jungle. However, as we evolved into complex societies, anger and aggression had to be tamed and rechanneled to serve us. But when needed, we have to be courageous and aggressive enough to enter into combat with an enemy. Aggression and anger seem to be intrinsic to all animal behavior. We have learned to suppress anger and refrain from overaggressiveness to form cultures and societies and to share and live together. Parental love and care, as well as religion, have contributed to molding our behavior in a fairly acceptable fashion. A problem may arise when we repress or deny our anger or aggression. We may not be aware that we are denying our anger or aggressiveness. Such behavior may have its origin at home—for example, when a mother tells her child to be a nice boy or a nice girl and not angrily tear apart a cheap toy the child's uncle brought home as a birthday present.

Looking for the root of repressed anger or guilt is not an easy task. After all, it has served us very well for many years. A worldwide goal for the near future is the sequencing of thousands of human genomes to identify many gene variants that will help us identify disease predisposition. Unrestrained aggression has been man's most destructive weapon. It may be considered a disease in need of restraining tools. Our DNA's revolutionary technology is fast advancing, giving us answers to questions we are working on now. Gene variations are secondary not only to changed base pairs but also to histone modifications through SNPs; this will surface as we unravel the genome's multiple component segments and their functions.

Michael, a psychotherapist with many years of experience, read the first part of this essay and suggested that modern DNA technology should develop appropriate tools to harness and modify the cluster or clusters of brain cells responsible for aggression. He maintains that those brain cells have lost their usefulness in modern society. He justifies his argument by pointing out the brain's ultimate weapons to defend itself from outside threats—atomic bombs. This is the ultimate tool of destruction. It will destroy not only human beings but also all other forms of life on planet Earth. I could not help but agree with him, adding that denial and greed are roadblocks on our peaceful journey.

You and I are aware that at present time, there are many health hazards that need to be addressed, but the vast majority of people choose to ignore this. Most people are aware that nicotine is a very strong causal agent in lung disease, but they continue to smoke despite the warnings. Drug abuse is another health threat present not only in the poor echelon of our societies but also among rich, powerful, and famous individuals. The most common argument by those individuals is that a knife can be used for defense as well as offense, which is a very poor analogy.

In the use of atomic weapons, almost four billion years of life development on planet Earth could be lost. Humans have spent most of their time on this planet fighting each other. Learning to look after and care for friends and neighbors is a very recent human attribute that has perhaps come out of life in towns and cities. We are not out of danger yet; slavery, discrimination, hunger, diseases, and abuse are rampant in many places on our beautiful and dear planet Earth. Michael's philosophical approach to our problems is sweet music to my ears, but I cannot afford to deviate from my original goal. I must return to the here-and-now issue: anger, guilt, and tears.

Anger is a nuisance in my brain that, although very useful during my long residence in the jungle and later on, while establishing towns and building fortresses for my protection, is now a liability. The worst part of it is that anger can turn inward, toward me. This problem seems to be aggravated when anger is repressed or denied. In that case, the human body's defense system is aroused, and otherwise unnecessary hormones flood the circulatory system, creating a hyperactive state of alertness that is harmful to many vital organs. Glucocorticoids—hormones from the adrenal cortex—are released in bundles in response to stressful experiences. Stress takes over when it need not do so. Once again, the ability to think, act, and behave rationally and intelligently is seriously compromised.

"In the hippocampus, a brain region densely populated with receptors for stress hormones, stress and glucocorticoids strongly inhibit adult neuro-genesis ... Decreased neuro-genesis has been implicated in the pathogenesis of anxiety and depression ..."[70] This same hormone that threatens neurogenesis may also save the life of an asthmatic person. These hormones are also potent anti-inflammatory medical tools that save life in all parts of the world. An excessive release of this hormone may provoke heart problems and strokes. It goes without mentioning that the amygdalae and prefrontal cortex are also in jeopardy when there is an overflow of this hormone.

It just occurred to me now that knowing all the negative ramifications of an overflow of this hormone may make me more susceptible to the release of this chemical in my body. However, it sounds as if I am saying that I should not explore any further assets or liabilities of glucocorticoids. These hormones have assisted me for too many years for me to abandon them now. In fact, I am inhaling a simple version of this hormone to control my asthma. Anger may provoke an asthma attack, but I did not experience any problem breathing normally the evening I saw the image of the little girl chained to a tree trunk. An asthma attack may be provoked by multiple causal factors or triggers, including medication for other maladies. We need to be taught when we are small children that a poisonous snake's bite or a wasp's honeybee's sting may be lethal.

---

[70] *Nature* (August 25, 2011).

I strongly dislike snakes, and I am not a friend of those other two flying creatures. I have been stung by both of those flying enemies of mine. I have no idea whether the tree trunk the girl was chained to had any connection with the trunk of the mango tree I was under when I was stung by a wasp. A Freudian analyst may insist on exploring this possibility, but as I informed you already, I want to proceed with a biological basis in investigating the anger I felt following my guilt and tears.

I feel I must clarify for my own peace of mind that major depression is considered by contemporary psychiatry and psychology to be a brain disease with a biological basis. During my early years of training, psychodynamics had the upper hand and Freudian theory considered major depression a disorder of the mind. The mind could not be subjected to medication while psychoanalysis was being employed. Patients used to spend many years on the couch with little progress, if any. It was an intellectual exercise with no personal affective interaction between patient and psychoanalyst.

Mary: From a Soft Couch to a Hard Chair

I RECALL A LADY—I WILL call her Mary—who spent years on a couch. Mary was a thirty-seven-year-old single Caucasian female working as a clerk at an airline office. Her only sibling, a boy two years her junior, died when he was seven years old. The cause of his death was leukemia. Mary lived with her parents, who took care of their home expenses. Her maternal uncle was the manager of the airline office. She had spent almost six years undergoing traditional psychoanalysis. Her high school years were described as normal, except that she did not actively socialize with her classmates. She had fairly good grades and decided to attend a local community college. Mary wanted to be a schoolteacher. During her first year in college, she did fairly well; but during the first semester of her second year, her work began to slip. Mary lost interest in her education, became easily upset without provocation, experienced crying spells and sleeping problems, easily tired, and, above all, lost interest in all social life. According to the chart handed over to me by the chief psychiatrist, Mary had had suicidal ideations and vague suicidal plans, but she had never been hospitalized. She had been raised in a religiously oriented home environment. She confessed her suicidal ideations to her priest. The priest told her that it is a major sin against God to consider self-destruction. He said that God had given her life and she must never go against God's plans. Mary had walked halfway across a bridge connecting Brooklyn and Manhattan and found a spot from which she could jump and kill herself. She said she did not do it because of the priest's warning to her. She felt she should not disobey God. During another occasion, she thought her mother's sleeping pills could be used to finish her unbearable pain and suffering.

Mary was referred to me by the chief psychiatrist in the hospital I used to work for. He was a European-born Freudian-trained professional, but he had an open mind. We used to have frank and professional arguments on orthodox psychoanalytic theory. I was a supervisor carrying almost a regular caseload. I took it upon myself to work with patients regardless of my position during my entire professional career. The chief psychiatrist was my immediate supervisor. His office door was always open to me. The day he called me into his office to assign Mary to me, he said, "It is all yours; she has had enough of the couch." He meant she was tired of traditional psychoanalytic theory and treatment.

My office was unadorned in order to avoid distraction of any kind; there was nothing on the walls except a calendar. The chair I was sitting on had a cushion to make my seven-hour day somewhat comfortable. The chair Mary was sitting on did not have a cushion; it was purposely arranged that way by me. Each therapeutic session lasted between forty and forty-five minutes. When I called Mary into my office and pointed to her seat, she kind of hesitated and looked around, seemingly looking for a couch or a softer seat. She briefly glanced at me, and then, looking at the floor, she said, "You are going to analyze me, are you not?"

Mary was neatly dressed; her facial makeup was lightly visible. Mary looked again at me with a half-forced smile, as if she were waiting for my answer to her opening question. I told her I had read her past treatment history and said that if any analysis had to be done, it would have to done by her. I added that any analysis done in my office had to be about something that was bothering her at the present time. I further advised Mary that my concern during our forty-five-minute session was her feelings as she was experiencing them at the moment. Without any other preparation, I asked how she was feeling while sitting on her chair. She seemed to be taken by surprise at my question and apologetically said she was all right. She proceeded to place her relatively long skirt under her legs without looking at me.

I said to her, "Can you look at me?" Immediately afterward, I added, "What I mean is, how do you feel about the way the chair you are sitting on is pressing against your body?" Again, she avoided getting into feelings and told me that she attended church services regularly and that the pews were not softer than the chair she was sitting on. Of course, Mary was not used to addressing her own feelings in the way I was asking her to. She was used to verbalizing whatever came to her mind; Her communications had been coming from the frontal cortex and left side of her brain. The limbic system, with all its emotions, had been left aside, waiting to be called in if necessary. It was a fragmented approach to repair a dysfunctional brain system.

Mary and I had not yet established a trusting working relationship in order for her to deal with her own self and the feelings that had provoked the symptoms that had caused her to come to our clinic for help. The establishment of a trusting working relationship between a patient and a therapist may take several sessions. There are several conditions that may determine how that relationship might take shape. Resistance to therapy may come in different shades and shapes. It is up to the therapist to lead and point out how to come out of a self-destructive state a patient may not be aware of. The therapist must temporarily donate his or her energy so the patient can crawl out of a very slippery emotional hole.

I explained to Mary that the association she had made between the church's pews and the chair she was sitting on in my office was an intelligent one. However, I explained to Mary that the chair she was sitting on was intended to allow her to feel her body as it related to her emotions and feelings. In an effort to make the session a little smoother, I looked at Mary and asked her to rest her arms on the chair arms. Rather nervously, Mary looked at me and said she was not used to looking straight at a therapist's eyes. She said her heart was palpitating faster than ever before. I said I was happy to hear that from her. I wondered if she felt her blood traveling all over her body. She paused for a little while but did not look at me. Mary then said she felt blood going to her head. Immediately after that, she turned her palms upward to show me she was sweating. Without saying a word, I conveyed to her I was supportive of her. She took a deep breath, and with what appeared to be a spontaneous smile, she said, "Thank you." This first session was over, and purposely, I said nothing else. I got up from my chair and opened the door for her to leave my office. When Mary had taken the last step out of my office, she turned around and said to me, "You did not tell me when I have to come back to see you again?"

Mary had gotten my message and wanted to further explore the emotions and feelings that had made her feel ill. She had felt her forearm against the chair, as well as blood flowing in her body. Equally beautiful for her was seeing a therapist in front of her sharing her fears and her emotions. There was nothing for me to interpret for her. She must have received more than enough interpretations during her prior treatment. What I wanted her to take home from our first encounter was that as an emotional human being, if she allowed me, we could join our energy and deal with the dreadful memories that had made her come to us for help. She did not have to talk about it over and over again, activating synapses and forming long-term memories. Memories that are already toxic need to be deleted, modified, or deactivated. Building new ones was my goal for Mary.

During the following sessions, Mary felt comfortable dealing with feelings as she experienced them during the therapy hour. I was there to assist her in getting back on track whenever an intruding thought provoked her to wander away from herself. I also intervened whenever an emotionally overloaded memory demanded extra energy to overcome a severely traumatic episode. Otherwise, Mary had to rely upon her own strength and resources when needed.

Mary was able to deal with her feelings toward her parents—particularly her mother. A religious lady, she told me she followed her faith's teachings dutifully. Mary also was very respectful of her parents' expectations for her. However, during our therapy sessions, she verbalized and acted upon feelings of anger toward her mother she had not shared with the priest during confession. During our sessions, she gained enough confidence in herself to describe her mother as a selfish and ignorant person. Later on, Mary changed her description, saying her mother was an overprotective mother who did not know any better at the time. On several occasions, Mary visualized her mother taking part in our sessions; she imagined her mother was sitting in front of her on top of a toy box I had in my office. Her impression of her mother fluctuated between a loving mother and a strict, ignorant person. She addressed her mother freely and without hesitation of any sort.

About five months into therapy, Mary asked me if she could come to see me twice a week. Without thinking about it, I told her it was nice to hear she wanted to come for therapy more often. I added that I would need to consult with my supervisor about it. In reality, I could not have made room for her even if I wanted to. Anyhow, I went to the chief psychiatrist, who pulled out a list of clients waiting to be called in for therapy. He said, "Take as many as you want." Then, in a joking manner, he added, "You are the only therapist asking me for more patients." After we had a good laugh at each other, he continued by adding, "I need you supervising all personnel in this clinic. I am busy writing proposals for this one beside our satellite clinic located on Rochester Street." I told him the name of the patient I was referring to. I told him I could accommodate her in my English-speaking group. I had another Spanish speaking group. He said to me, "Do as you please, but do not forget what I said about your main job here."

I informed Mary about our decision, and she looked very happy. In fact, she had already talked with two ladies from the group who had welcomed her decision. Mary was coming out of her self-built shell and beginning to enjoy the fruits socialization had waiting for her. Later on I learned that some group members occasionally went out together to museums and Broadway shows. Before Mary joined the group, I explained to her that the group existed as if it had a personality of its own. During each group session, she was not to refer to herself as an individual unit existing outside the group. She had to fuse with the group, forming a larger collective unit.

During one of our individual sessions, Mary volunteered to work on a dream that was provoking anxiety in her. The dream turned out to be related to her nephew's attempt to play with her genitalia. Her nephew had been invited by her mother to spend weekends at home. Mary was aware of my toy box and its contents, which were available for use during each session. While working on the dream, Mary went straight to the toy box, pulled out a dummy the size of her nephew, and placed it on her seat. She then went back to the toy box, pulled out a rubber baseball bat, and began hitting the dummy, calling her nephew by his name. Fatigued and looking tired, Mary looked at me for approval. I nodded. She pulled the dummy out of her chair and threw it on the floor, adding in a loud voice, "You do not deserve to be sitting on my chair; get out of my head and my life." She celebrated the session, saying, "I have wanted to do this for a long time. I am glad I could do it today. Thank you."

Mary had shed many tears as a child and as an adult. She was also loaded with feelings of guilt that had driven her to consider terminating her life. She needed more than verbal therapy; she needed to take action against toxic memories that had invaded her brain. Mary continued to attend individual and group therapy until I was promoted to another job in another clinic seven months later.

About a year later, my daughter and I went to see a ballet play in central Manhattan, not far from Broadway. We were on the second floor, looking down at the nearby plaza. We were early for the show. To my surprise, an elegant, beautiful, and well-dressed lady came running toward me. Without saying a word, she hugged me and kissed me on both cheeks. Right behind her was a handsome and a sharply dressed man who appeared to be about forty years old. It took me a little while to come to my senses after such a display of affection from a lady who had wanted to die a few months before. She turned around and said, "Tom, this is the therapist I have talked to you about. I want to introduce you to him." While I was shaking Tom's hand, Mary said to me, "Tom is my husband." You can imagine the feeling of joy that spread all over my body. Seeing the fruits of my concern and love for my neighbors is overwhelming. My search for health and happiness makes me feel extremely happy when I enjoying seeing someone else's success. Not all my cases had happy endings like the one Mary experienced.

Although not a single person undergoing therapy with me ever committed suicide, some required emergency hospitalization. However, I recall a patient of mine—I will refer to him as Tito— who came from the same town that I was born in. He had been incorrectly diagnosed as schizophrenic. His command of the English language was limited, and at the time, Spanish-speaking psychiatrists and psychotherapists were few. In our clinic, we diagnosed him as having bipolar disorder, manic. He was placed on appropriate medication and showed remarkable behavioral improvement. However, improvement for him meant that he was cured, and so he stopped taking his medication.

During a visit to my parents, Tito's mother, who was a friend of my mother, pleaded me to help her son. I promised her I would do everything possible to help him. He had no relatives living in New York City, and his friends were very few. Luckily my mother had given his mother my home phone number. It did not take long for Tito to get my home telephone number. During his manic stages, he used to go traveling across the United States. I recall him calling me from Philadelphia, Miami, and San Antonio, Texas. His call from Philadelphia was easily taken care of. I lived in Philadelphia for four years. In Miami I had relatives and friends who were friends of Tito from childhood. However, they preferred for Tito to stay away from their homes and families. I was able to arrange a bus ticket for Tito to get to New York City.

The funniest call from Tito came from San Antonio, Texas. I had made a visit to that city a couple of years earlier and remembered some important landmarks. He called me saying he was very tired, sleepy, hungry, and without a penny in his pockets. When I asked him where he was calling from, he said, "Texas."

I said, "Thank you, Tito, but Texas is the second largest state of the Union." When I asked him who was he with, he handed over the telephone to a Mexican lady. She told me it was San Antonio. At the time, there was an agency called Travelers' Aid. I knew the location of that agency and directed Tito there. I contacted the agency, and they agreed to provide Tito with a one-way ticket to New York City after I promised to refund them from our clinic.

I have met some very interesting individuals during my career as a professional psychotherapist. Brain disorders and brain dynamics have been my concern for about forty-five years. Whenever I had a patient in front of me during a therapy session, I could not help but imagine his or her brain looking like a Christmas tree covered with many lights that were turning on and off many times per second. The limbic system, including the prefrontal cortex, followed by the association areas of the brain were my favorite regions to focus on during the therapy hour. Occasionally I used to say to myself, "Neurons, wake up and get this man in action, help him overcome this impasse." I wanted the emotion to be unlocked so that the patient could use all of the tools at her or his disposal to weaken the strength of the toxic memory at hand.

Helga, the Star of the Circus

I HAVE TALKED ABOUT DNA, RNA, proteins, the genome, and epigenetics, followed by the little girl chained to a tree trunk, Jeanne, Mary, and Tito. You should not have any problems identifying or dealing with any of the above subjects. The molecules DNA and RNA, proteins, and the genome are so familiar to me that even during breakfast they come in different letters, shapes, and colors. The patients I referred to are people you might come across at any time while walking on the street. I am trying to put all of the above in one solid picture, if possible. I am trying to form molecules, proteins, cells, and diseases into a living picture capable of conveying all the problems people confront and overcome during their lifetimes. I know this is difficult, but it is not impossible. If I fail in my attempt, there is someone out there who is capable of doing it. I have talked about the brain, brain diseases, and disorders, including the victim's pain and suffering when the brain becomes dysfunctional. In retrospect, I believe that I have left out a very important component of the brain puzzle—in particular, the neurotransmitter dopamine. This neurotransmitter is implicated in Parkinson's disease, Alzheimer's, schizophrenia, depression, and addiction, besides being the "feel-good" king of all brain chemicals. You already know that the largest number of dopaminergic cells is located in the substantia nigra in the pons. I also mentioned to you that the ventral tegmental area and the nucleus accumbens are the main activation centers during drug addiction. I did not forget endorphins (endogenous morphine); I will address these briefly if time and space permit me to.

I remember a patient in my clinic—I will refer to her as Helga—who seems to me to be another good illustration of the brain, brain diseases, biochemicals, behavior, medication, and psychotherapy. Helga's parents were of Northern European ancestry and had migrated to the United States. Her father, John, was a truck driver, and her mother, Karen, was a waitress at a large and busy restaurant when they met for the first time.

Helga and her parents lived next to Karen's mother's home. They rented a small but comfortable house near Helga's grandmother. Helga's father, John, had to stay away from home overnight when he had to make deliveries a long distance from home. When Helga was five years old, her mother died. Helga's maternal grandfather had died when she was three and a half years old. When Helga became an orphan, she was taken care of by her grandmother for a few months. Her grandmother was diagnosed with lung cancer that soon metastasized. Her father put Helga up for adoption through a local religious organization.

For the next ten years, Helga seemed to have made adequate adjustments under the care of her adoptive parents. She maintained sporadic contact with her putative father. However, after Helga reached her sixteenth birthday, she began to have provocative arguments at home and at school. Three weeks before her adoptive parents were ready to celebrate her seventeenth birthday, Helga left home. Her adoptive parents and her father made all attempts possible to find Helga and bring her home. They were told that Helga was seen at a local gas station carrying a small backpack. She had gotten into a car with two boys and another young girl. Unknown to her father and adoptive parents, Helga had joined a small circus that went from town to town with little success. The main feature on the circus was acrobatic stunts like performing trapeze tricks and walking the tightrope. There were no elephants, smart chimpanzees, or talking birds to amuse the children. Helga had become very good at jumping from one trapeze bar to another and enjoyed her job a lot. Helga had become the main attraction with her daring and spectacular jumps. She was so happy being the main attraction of the show that she contacted her father and adoptive parents. Her father offered to open a restaurant in a nearby town and let her run it. However, the offer was not appealing to Helga. She was the star of the circus, and cheers made her feel like a princess.

Helga had had a rough time during the death of her mother and grandparents. She had caring and loving adoptive parents, but the emotional scars left on her by the death in her family were problematic for her. The cheers and long ovations she was getting from the circus crowd were, in a way, compensating for the emotional vacuum that had provoked her to run away in the first place. For her, the circus job was more than a simple job; it was a resource for emotional feedback she did not want to miss out on. Besides performing on the trapeze, Helga had developed a set of daring numbers that made the public call for Helga over and over again after each performance. Money was coming to the circus in a way no one had dreamed of before.

The circus manager decided to bring another girl into the show. The new girl, Molly, was not meant as competition for Helga on the trapeze, but Molly's beauty, charisma, and magic tricks (in which she pulled items out of both her hat and her brassiere) made Molly's act Helga's most hated act in the circus. She began to blame and accuse the new girl and the manager of plotting against her. It seemed to Helga that the guardian angel that had protected her was abandoning her for the new girl. And as with all disasters in life, the worst takes place when you are least prepared to overcome it.

Helga wanted to outshine the new girl by making jumps she had not done before. Each time Helga performed a sensational feat, the public stood up to cheer for her audacious number. Actually, Helga was not competing against Molly. There was nothing going on between the manager and Molly. Helga was fighting her own delusions; she was competing against herself. And when you fight against yourself and you are not aware of it, your body is the battleground. Even her manager cautioned her against performing some of her feats, but that only reinforced her suspiciousness toward him.

Soon, the last day for Helga's performance in the circus arrived. It was a Saturday evening after she had had an argument with Molly. There was anger reinforced by more adrenaline running through her blood, invading every cell in her body and brain. Dopamine, the feel-good neurotransmitter, had gone underground, and adrenaline was in command. Logic and rational thinking attributes normally assigned to the frontal cortex ran for cover at the sight of the avalanche of adrenaline. The locus coeruleus nucleus in the brain stem was pumping out its neurotransmitter without ceasing. The old survival nuclei had taken over Helga's neocortex. Not seriously considering the risk she was taking, Helga's mind was focused on outperforming the new girl. Actually, she was trying to outperform her own delusional self.

There is no doubt her judgment was seriously compromised at the time when she attempted to do a rollover in the air when jumping from one trapeze bar to another. Helga missed the trapeze and hit a pole, fracturing her right arm in two places. She was rushed to the local hospital, where surgery was performed on her. After a week of hospitalization, Helga was discharged to her father's home. Not only were her days as a circus star over, but her mind's delusions had begun to take over her brain on a grand scale as well.

When Helga came to our clinic, she had been hospitalized three times for psychiatric problems. Twice she had attempted suicide by combining alcohol and an overdose of her own medication. Her father used to bring alcohol home and lock it in a chest; he then carried the key for the chest in his pant pocket. However, it did not take long for Helga to outsmart her father—especially after he had a couple of drinks and went to sleep.

Helga's last hospitalization in a public city hospital was scary and very painful for her. She was brought in by the police on charges that she was soliciting money for sexual favors. The psychiatric hospital she was admitted to was awfully overcrowded. During the day, Helga was thrown into a psychiatric ward with sixteen other persons of different ages who had mental disorders. There was no separation of men and women. Most, if not all, psychiatric patients in the hospital were city residents. Some had to be restrained and taken out to be put in isolation because of overt aggressiveness toward other patients—particularly out-of-state females. Helga was basically a country girl who had been raised under strict religious observance until she joined the circus. Helga was treated roughly by other psychiatric patients. In addition to having a brain disease, she was ill prepared to deal with poor, uneducated, and mentally ill city dwellers. For some patients, a clean bed and three meals a day in the hospital was the best they had ever had.

Helga begged her father to help rent a studio apartment for her to get adequate and necessary psychiatric treatment in the city. She had been on several medications, including, but not limited to, haloperidol. Helga used to get monthly medication refills without follow-up psychotherapy. Several times she just stopped taking her medication, and it was then up to her father to get her back on medication. Her father was a caring person, but talking and listening to his daughter's stories was not his best attribute. His hobby was collecting stamps, particularly those from England, Germany, and Russia. Both Helga and her father used herbs, vitamins, and minerals as part of and in addition to prescribed medications. Carrots cut in small pieces, salvia, saffron, chamomile, and blackberries had magic components only Helga and her father knew about. Both took large doses of niacinamide "to keep their mind healthy." While in therapy with us, Helga joined a religious group that kept strict religious rules of behavior and engaged in daily reading and street proselytizing.

When Helga was not quoting the Bible, she was deeply involved in her aristocratic lineage. Helga's delusions were very real to her, and she would share them with anybody. She said there were people plotting against her to take away her rightful place in the royal families of Europe. According to her story, she was connected by blood to Alexandra, the Romanov czarina executed by the Bolsheviks in 1917. She also claimed to be connected to the English royal family by Queen Elizabeth's husband. Her life history had to be kept in strict secrecy because Bolsheviks and their sympathizers, all fanatic atheists, were always plotting against her. A secret she begged me not to reveal to anyone under any circumstance was that her mother was the daughter of Stacy. Stacy was the secret name of her grandmother, Anastasia, outside Russia. That made her a granddaughter of the late Russian emperor Nicholas II. But her delusions did not stop there.

Helga believed she was related by blood to Queen Nofry, short for Nefertiti, wife of Akhenaten. Further hiding her own self-identity, Helga claimed her past ancestors came from another constellation called Aldeb, short for Aldebaran. They came ten thousand years ago to teach Earth's humanoids. Helga, with the support of her father, kept a notebook with numbers and geometrical figures she insisted had been passed down to her in great secrecy by her ancestors thousands of years ago. Among her favorite numbers were five, three, and seven. In second place, but not less important, were one, nine and six. She mentioned and used the number five in her book more often than any other number. She tended to reduce names of people and places to five letters. Helga revealed this secret under the condition that I would never reveal her identity. According to her, her life history can be found in the Apocalypse of John.

She assigned the number five to words like blood, water, spade, cross, Jesus, souls, sky, stars, light, etc., all in consonance with her five-letter name. For her, the number one is the creator; it is alpha and omega. The numbers six and nine are different stages of the same person in his or her evolutionary trip in the universe. While a person is number six, he or she is provoked by multiple choices leading to and opposite of each other. The number nine is the hardest developmental stage to achieve. From stage nine, you can fuse with number one. She had three notebooks full of her stories. She shared with me only a small fraction of her life story, because she could read my mind. I was a person she could trust, she said. She began to trust me when she saw an orthodox Christian church cross on the cover of a book on my desk in my office. Helga claimed she had Alexandra's orthodox cross at her home. During one session, she said that I looked like an orthodox priest and that she could therefore trust me with her blessed cross. On another occasion, Helga made me a member of the group of aliens that had come to teach Earth's humanoids.

My therapeutic sessions with Helga were basically supportive of her efforts to help herself through socialization and reading religious and self-help books. I gave her clear, direct, and focused homework to do. I asked her to use the local public library dictionaries to find the meanings of the words she used most frequently to justify her own delusions. I reinforced this behavioral change by asking her to get feedback from members of the religious group she belonged to as well as from friends and people on the street. In other words, I placed on her the responsibility for checking out the validity of her false and distorted thoughts and ideas. I do not fight the victim; I help the patient to confront his or her dysfunctional self. The patients themselves make appropriate and healthy brain connections.

Helga was an intelligent lady. She enjoyed reading books other than those provided by her church group. Less fortunate men and women who could not read as well as Helga could were happy listening to her read for them. In a way, Helga became a star again. Later on, she was registered in a city-sponsored vocational training program leading to a secretarial job. Helga's diagnosis was never clear to me. There were periods of time when she was a lucid and well-behaved lady who engaged in logical and rational conversation. Anyone not familiar with Helga's life history would not suspect her of suffering psychosis followed by ideas of grandiosity and persecution.

Helga's psychiatric chart contained three different diagnostic opinions from different treatment agencies: bipolar disorder, generalized anxiety disorder, and schizophrenia. You may wonder why Helga had three different psychiatric diagnoses. In the absence of a definite biological marker for schizophrenia and bipolar disorder, manic phase, a definite diagnosis may be hard to arrive at. This confusion is more common when patients are taken to a psychiatric emergency room. Firstly, psychiatrists in training practice there. Secondly, a diagnosis has to be written from scratch. In the majority of cases, patients' past records are not available at the time and place of the diagnostic interview. Most often, these patients are brought to the emergency room by ambulance or police car. The input of their history into the diagnostic impression is very limited.

You may be wondering why Helga had been placed on various medications. The medications were prescribed to deal with her symptoms at the time of each diagnostic impression. She self-medicated with over-the-counter drugs and supplements. Both Helga and her father referred to books on herbal remedies. Another brain and bodily input was her change of food intake. We must remember the brain's capacity to change and improve plasticity—the taking over of functions not performed before. We grow new neurons with the potential for new synapses and new memories. We are not entirely hardwired.

I spent the largest portion of this essay talking about DNA mutations, genome variants, and protein genesis. I did not ignore epigenetics and environment, both of which shaped the way Helga felt, thought, and lived. She shed tears, became angry, and felt guilt, but she did not give up and kept on track, trying to overcome a brain disease.

Dysfunctional brain cell circuits and pathways influenced by variant causal agents may be brought back to an acceptable equilibrium of brain chemistry and behavior, thus improving a patient's quality of life. Sour, toxic, and chronically stressful life experiences can kill brain cells, especially in extremely sensitive areas, such as the hippocampi, amygdale, and prefrontal cortex. They are extremely important centers of cells implicated in learning and memory formation. They are risk areas in predisposed individuals. Therapists need to identify basic clinical symptoms with great accuracy, in the absence of biological markers, in order to begin administering preventive medicine as early as possible. Of course, we also need the patient's cooperation.

Helga's erratic behavior at the circus was not easy to stop. She was forcibly driven by chemical impulses impairing her capacity for logical and rational thinking. School education, as well as television documentaries and shows aimed at adolescents and parents, would be a good start. The issue is not only the pain suffered by patients and their families; it is also the billions of dollars spent attempting to care for them. This is money that could go toward research, prevention, and science-based treatment. I emphasized to my patients the need to prevent relapses and diseases in general.

We do not have to have correct answers for all questions. We often feel embarrassed when we cannot come up with a satisfactory answer. We spend a lot of energy and time trying to please the public at our emotional expense. This type of behavior is self-destructive and leads to illness. Prevention is very helpful on many occasions. Good planning is essential to prevention. You do not need to be as rigid as an iron rod and plan your day before you encounter stressful conditions. Your brain guides you to accomplish your goals while you make necessary concessions. Before you leave for your job or go on a trip, reassure yourself that you are an intelligent and healthy person capable of making logical and rational decisions.

You can choose when to enter a conversation as well as when to walk away from a poisonous and unhappy individual. You have the power to do so during almost any situation. There are individuals whose brains are wired in a mixed-up fashion. You might hear of college students, professors, legislators, law enforcement officers, and many others committing horrifying crimes you would have never have suspected them of. You have the power and the wisdom to walk away from seriously compromising situations before they harm you.

You can make a list of stressful situations you may encounter during the day and prepare yourself for multiple possible appropriate solutions; this leaves you in command during any given situation. Remember: by planning for problems ahead of time, you prepare your brain to come up with the solutions that will serve you best. Your brain has solved myriad problems you may have considered extremely difficult. You have billions of brain cells making trillions of connections to serve your needs. Do not attach yourself to a limited number of objects at your disposal. When you attach yourself to one objective, as in the case of Helga, you miss all the opportunities in front of you. Individuals attach themselves to cars, jewels, televisions, alcohol, and drugs. Being greedy for the sake of power seems to be the greatest sin you can commit against yourself.

I should not close out these good-health practices without mentioning once more how the brain gets hooked on pleasure-seeking behavior. I mentioned dopamine and endorphins as brain chemicals that may determine whether you get stuck on a repetitive self-destructive behavior as opposed to choosing good health and happiness. Once you allow those two chemicals to hijack your brain, you have given up all chances of becoming a happy person. Your ventral tegmental area and nucleus accumbens will light up like a Christmas tree and stay on until they destroy you.

Helga and Mary needed to deprogram themselves and leave behind behavioral responses that prevented them from enjoying health and happiness. Toxic and painful memories that accumulate in your brain from birth have to be modified or removed from your brain. Helga's delusion that she was a princess with royal blood from Europe's two most powerful dynasties blocked her realistic thinking and good judgment. Despite her risk-taking job of performing on the trapeze, her mind was wandering in faraway lands, bound to emotional attachments. Even if there were a genetic component in Helga's brain breakdown, preventive therapeutic tools could have prevented stressful and painful episodes.

As we grow up, we make many unnecessary attachments at the expense of our inborn and experimental survival tools we carry from our ancestors' past experiences. In today's faddish lifestyle, we overextend our resources, becoming embroiled by unnecessary stressful conditions. We go against our brain's powerful resources and rely on childish tricks for survival. Helga and many others like her are in need of skillful professional therapeutic intervention to tear down toxic brain barriers and prepare them for good health and happiness. It may be hard to believe, but they need to establish new neuronal circuits and networks. Past abnormal behavior seems to be in front of their thoughts and actions, preventing them from focusing their attention in a given present situation. Toxic, painful memories bring about structural changes in dendrites. Those memories attempt to stay active and monopolize our attention and thoughts. By analogy, toxic painful memories are like parasites that feed and grow, attracting healthy thoughts and memories.

During my initial therapeutic sessions with Helga, she could hardly stay with me in focusing her attention on a specific feeling or thought. Her brain had stored many confusing memories from her alleged royal family, and those memories were entangled with cryptic messages. As far as I was concerned, I was talking to an impostor who had taken over Helga's brain. The lady sitting in front of me giving me cryptic messages based on numbers and words gave me the impression that she was not in my office. Her body was sitting in front of me, but mentally and emotionally, Helga was in a faraway land. It is not that she did not care about herself and her treatment with me. She was always on time and did not miss a single session with me. Helga took refuge from daily life challenges by building an imaginary family. To keep up with her thoughts of exaggerated importance, she chose royal families and cryptic messages. By escaping into these mental fantasies, she kept her brain busy.

Helga may have found respite from everyday challenges and needs by escaping into her imagined palatial life. In a way, she was replacing past toxic memories with memories she chose to enjoy. However, there was a big problem to be overcome during working sessions. Escaping to her world of fantasy brought Helga relief from immediate demands around her. She dressed, talked, and attempted to behave like royal princesses do. Her behavior at the clinic caught the attention of our staff. They referred to Helga as my aristocratic friend. My immediate task was to bring Helga to feel her body on a rough chair. I could not deprive Helga of her runaway trips to Wonderland. I had to guide her to find and identify her own strengths and attributes, which she was placing on a fantasized princess. She had to find herself beautiful, intelligent, educated, charming, and worthy of public acceptance and praise for her talents.

Gaining Helga's trust and cooperation was paramount for therapeutic success. I needed an object she could see and touch during our sessions. Inside my toy box, I had a game that looked like a Chinese abacus. It had three horizontal wooden rods holding nine beads. It could stand by itself on top of my desk. I introduced it as a possible therapeutic tool during our third therapy hour. Helga could not contain her curiosity about the toy. She asked me how the game was played. I explained to her that I was going to test us on how long we could stay focused on a stated subject or theme. If we were talking about her feelings while she was putting on an imagined princess's dress and she jumped to another theme, we would mark it by pulling a bead to the opposite side. I placed emphasis on *we* as opposed to *I* or *you*. When she jumped from "I feel sad now" to another topic, such as her father's stamp collection, I made a hand signal to stop her and moved a bead to another position. Every time I moved a bead, it meant we were not doing anything compatible with psychotherapy.

It was during the following session that Helga said, "I have to be careful how I talk; I will stay here, as you want me to." I managed to get Helga to tell me she had changed our rules to stay on track. This was the beginning of her initiation of self-guided corrective thoughts. Helga was making progress in psychotherapy other than taking her medication as prescribed by our psychiatrist. Helga was beginning to make changes in her behavior. She began to call the Chinese abacus her brain's inner clock. Her mind was transferring attention from her imagined royal family connections to present conversations with me.

My immediate goal was for Helga to stay with her feelings and actually feel herself as a unique person. She needed to feel the real Helga sitting in front of me. She had to refer to herself in first person singular: "I feel relaxed," "I am here with you now." I interpreted this as a way of her making new synaptic connections.

The Chinese abacus, her brain clock, and the rules she accepted during our sessions began to take her away from her imagined overseas trips. When I did not place the abacus on my desk, Helga would go to my toy box, pull it out, show it to me, and place it in front of us. To my surprise, one day she came to my office carrying a new and bigger Chinese abacus. She said, "This one is for you." Helga's therapy sessions and medication were contributing to the rediscovery of a healthy and happy young lady.

Take-home work was part of our treatment contract. She had to get feedback from family members and friends as to how she was progressing in our treatment program. Her father was a very supportive person. He used to call her a very intelligent person. During therapy we would go over her compliments. Helga would say aloud, "I am an intelligent lady." She had to repeat it several times over. She liked to repeat, "I have a beautiful ballerina body." On several occasions, Helga compared me to her father. She felt she did not have to compete against anybody. My demand on her was to stay focused on her body. Emphasis was placed on her body as opposed to her role in an imagined royal house. In our sessions, Helga did not have Molly to compete against or an audience demanding more of her talent. Most important, she did not have to compete against herself. Self-competition was replaced by self-realization, self-identity and self-accomplishments. Predisposition to heritable brain diseases and problems is not easily overcome. Early identification and professional guidance and intervention are the best tools we have. They can prevent a lot of tears, guilt, anger, and pain. We do not need to see children chained to a tree trunk. We do not need to hear that a mother has drowned her children or that an angry adolescent boy shot and killed his parents. Education and prevention are blessings we have received; they are tools we need to apply before a tragedy strikes us.

Please, get an image of a human brain divided into right and left hemisphere. Focus your attention on its internal groups of cells and their functions. Identify the frontal cortex and the cerebellum on the back of your brain. Look for the hypothalamus, thalamus, amygdalae, hippocampi, and brain stem. Think of it a game—a game with your brain. It is not only a therapeutic tool of mine to keep troubling thoughts away; it is also an excellent educational tool. Think about the damage you could do to a cluster of brain cells if you became hooked on a destructive behavior; this brain game could help you stay healthy.

DNA Double Helix

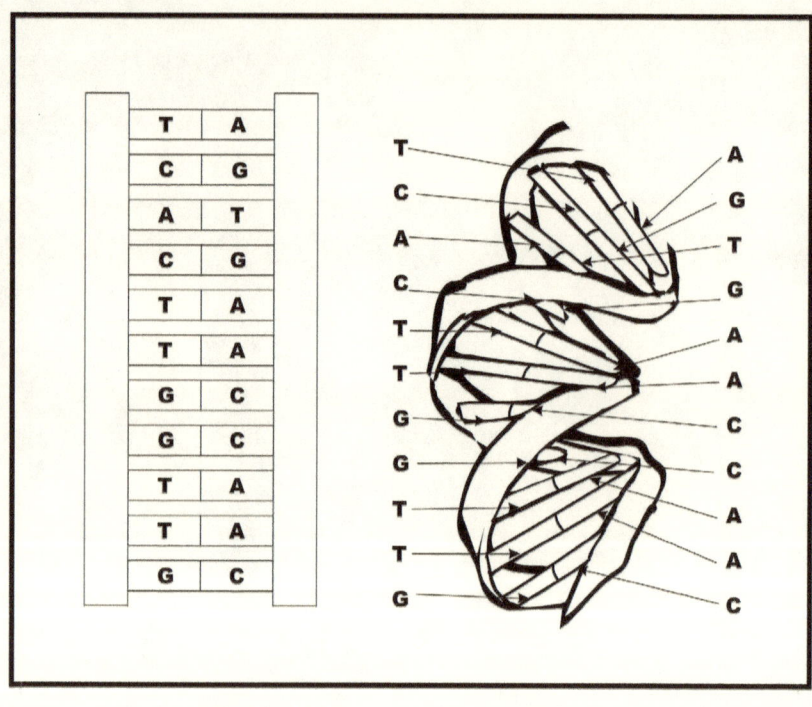

www.ingramcontent.com/pod-product-compliance
Lightning Source LLC
Chambersburg PA
CBHW020743180526
45163CB00001B/326